Human Reliability, Error, *and* Human Factors *in* Engineering Maintenance

with Reference to Aviation and Power Generation

Human Reliability, Error, *and* Human Factors *in* Engineering Maintenance

with Reference to Aviation and Power Generation

B.S. Dhillon

CRC Press
Taylor & Francis Group
Boca Raton London New York

CRC Press is an imprint of the
Taylor & Francis Group, an **informa** business

CRC Press
Taylor & Francis Group
6000 Broken Sound Parkway NW, Suite 300
Boca Raton, FL 33487-2742

Library of Congress Cataloging-in-Publication Data

Dhillon, B. S.
    Human reliability, error, and human factors in engineering maintenance : with reference to aviation and power generation / B.S. Dhillon.
        p. cm.
   "A CRC title."
   Includes bibliographical references and index.
   ISBN 978-1-4398-0383-7 (hardcover : alk. paper)
    1. Human engineering. 2. Human-machine systems. 3. Errors--Prevention. 4. Reliability (Engineering) 5. Fallibility. 6. Airplanes--Maintenance and repair--Quality control. 7. Electric power plants--Maintenance and repair--Quality control. I. Title.

TA167.D468 2009
620'.0046--dc22
                                             2009005529

Visit the Taylor & Francis Web site at
http://www.taylorandfrancis.com

and the CRC Press Web site at
http://www.crcpress.com

Dedication

This book is affectionately dedicated to all my schoolteachers, including Mr. C. Bell and Mr. G. B. Gill at the Dale School for Boys, Derby, England, for their inspirational comments and constant encouragement.

Contents

Preface

Each year billions of dollars are spent to maintain engineering systems throughout the world. For example, U.S. industry spends over $300 billion on plant maintenance and operation annually. It is estimated that about 80 percent of this amount is spent to rectify the chronic failure of systems, machines, and humans.

Over the years, the occurrence of human errors in maintenance activities has been following an upward trend due to various factors, and the resulting consequences can be very serious. Two examples of these consequences are the Three Mile Island nuclear accident and the crash of a DC-10 aircraft at O'Hare Airport in Chicago.

Over the years, a large number of journal and conference proceedings articles on human reliability, error, and human factors in engineering maintenance have appeared, but to the best of this author's knowledge, there is no book that covers these three topics and includes maintenance safety within its framework. This causes a great deal of difficulty for engineering maintenance professionals because they have to consult many different and diverse sources.

Thus, the main objective of this book is to combine these topics into a single volume and eliminate the need to consult many diverse sources in obtaining desired information. The sources of most of the material presented are listed in the reference section at the end of each chapter. These will be useful to readers if they desire to delve more deeply into a specific area or topic of interest.

The book contains a chapter on mathematical concepts and another chapter on introductory material to human factors, reliability, and error, which are useful for understanding materials presented in subsequent chapters. Furthermore, another chapter is devoted to methods considered useful for performing human reliability and error analysis in engineering maintenance.

The topics covered in the book are treated in such a manner that the reader will require no previous knowledge to understand the contents. At appropriate places the book contains examples along with their solutions, and at the end of each chapter there are numerous problems to test the reader's comprehension. An extensive list of publications dating from 1929 to 2007, directly or indirectly on human reliability, error, and human factors in engineering maintenance, is provided at the end of this book to give readers a view of the intensity of developments in the area.

This book is composed of 11 chapters. Chapter 1 presents historical developments in human factors, human reliability and error, and engineering maintenance; important human reliability, error, and human factors in engineering maintenance–related facts, figures, terms, and definitions; and sources for obtaining useful information on human reliability, error, and human factors in engineering maintenance.

Chapter 2 reviews mathematical concepts considered useful to understanding subsequent chapters. Some of the topics covered in the chapter are Boolean algebra, probability properties, probability distributions, and useful definitions. Chapter 3 presents various introductory human factors, reliability, and error concepts.

Chapter 4 presents a total of eight methods considered useful for performing human reliability and error analysis in engineering maintenance. These methods are failure modes and effect analysis, man–machine systems analysis, root cause analysis, error-cause removal program, the cause-and-effect diagram, the probability tree method, fault tree analysis, and the Markov method. Chapter 5 is devoted to human error in maintenance. Some of the topics covered in this chapter are the maintenance environment, causes for the occurrence of maintenance errors, types of maintenance errors, typical maintenance errors, and useful design improvement guidelines to reduce equipment maintenance errors.

Chapters 6 and 7 present various important aspects of human factors in aviation maintenance and power plant maintenance, respectively. Chapter 8 is devoted to human error in aviation maintenance. It covers topics such as human error occurrence causes in aviation maintenance, types of human errors in aircraft maintenance, common human errors in aircraft maintenance activities, maintenance error decision aid (MEDA), and useful guidelines for reducing human error in aircraft maintenance activities.

Chapter 9 presents various important aspects of human error in power plant maintenance, including facts and figures, causes of human error in power plant maintenance, maintenance tasks most susceptible to human error in power generation, and steps for improving maintenance procedures in power generation. Chapter 10 is devoted to safety in engineering maintenance. Some of the topics covered in the chapter are facts and figures, maintenance safety problem causes, factors influencing safety behavior and safety culture in maintenance personnel, and guidelines for engineering equipment designers to improve safety in maintenance.

Finally, Chapter 11 presents a total of seven mathematical models for performing human reliability and error analysis in engineering maintenance.

The book will be useful to many individuals, including engineering professionals working in the area of engineering maintenance; maintenance engineering administrators; engineering undergraduate and graduate students; maintenance engineering researchers and instructors; maintainability, safety, human factors, and psychology professionals; and design engineers and associated engineering professionals.

The author is deeply indebted to many individuals, including friends, colleagues, and students for their invisible input. I thank my children, Jasmine and Mark, for their patience and intermittent disturbances that resulted in many coffee breaks! Last, but not least, I thank my wife, Rosy, my other half and friend, for typing various portions of this book and for her timely help in proofreading.

B.S. Dhillon
Ottawa, Ontario

Author Biography

Dr. B. S. Dhillon is a professor of engineering management in the Department of Mechanical Engineering at the University of Ottawa. He has served as chairman/director of the Mechanical Engineering Department/Engineering Management Program for over ten years at the same institution. He has published over 340 articles (199 journal and 141 conference proceedings) on reliability, safety, and engineering management. He is or has been on the editorial boards of nine international scientific journals. In addition, Dr. Dhillon has written thirty-four books on various aspects of reliability, design, safety, quality, and engineering management published by Wiley (1981), Van Nostrand (1982), Butterworth (1983), Marcel Dekker (1984), Pergamon (1986), and so on. His books are being used in over 85 countries, and many of them have been translated into languages such as German, Russian, and Chinese. He served as general chairman of two international conferences on reliability and quality control held in Los Angeles and Paris in 1987.

Professor Dhillon has served as a consultant to various organizations and bodies and has many years of experience in the industrial sector. At the University of Ottawa, he has been teaching reliability, quality, engineering management, design, and related areas for over 29 years. He has also lectured in over 50 countries, including giving keynote addresses at various international scientific conferences held in North America, Europe, Asia, and Africa. In March 2004, Dr. Dhillon was a distinguished speaker at the Conference/Workshop on Surgical Errors (sponsored by the White House Health and Safety Committee and the Pentagon), held on Capitol Hill (1 Constitution Avenue, Washington, D.C.).

Professor Dhillon attended the University of Wales, where he received a BS in electrical and electronic engineering and an MS in mechanical engineering. He received a PhD in industrial engineering from the University of Windsor.

1 Introduction

1.1 BACKGROUND

Since the Industrial Revolution, maintenance of engineering equipment in the field environment has been a challenging issue. Although over the years impressive progress has been made in maintaining equipment in the field environment, maintenance of equipment is still a challenging issue with regard to factors such as cost, complexity, competition, and size. Each year billions of dollars are spent to maintain engineering systems throughout the world. For example, U.S. industry spends over $300 billion on plant maintenance and operation annually [1]. It is estimated that about 80% of this amount is spent to rectify the chronic failure of systems, machines, and humans.

Over the years, the occurrence of human errors in maintenance activity has been following an upward trend due to various factors, and their (human errors) consequences could be very serious. Two examples of these consequences are the Three Mile Island nuclear accident and the crash of a DC-10 aircraft at O'Hare Airport in Chicago that killed 272 people on board [2–5]. Since the late 1920s, a large number of publications directly or indirectly related to human reliability, error, or human factors in engineering maintenance have appeared. A list of over 200 such publications is provided in the Appendix.

1.2 HISTORY

This section presents an overview of historical developments in human factors, human reliability and error, and engineering maintenance.

1.2.1 HUMAN FACTORS

The history of human factors may be traced back to 1898, when Frederick W. Taylor conducted various studies to determine the most appropriate designs for shovels [6]. In 1911, Frank B. Gilbreth studied bricklaying, and that resulted in the invention of a scaffold. As the result of this invention, the number of bricks laid per hour by bricklayers almost tripled (i.e., from 120 to 350 bricks per hour).

In 1918, the United States government established laboratories at the Wright-Patterson Air Force Base and the Brooks Air Force Base for conducting human factors-related research [7]. The period between World War I and World War II witnessed significant growth in disciplines such as industrial engineering and industrial psychology. By 1945, human factors engineering was recognized as a specialized discipline of engineering. In the 1950s and 1960s, the United States military and space programs further increased the importance of human factors in system design.

Currently, there are hundreds of published documents on various aspects of human factors in the form of textbooks, technical reports, design specifications, and articles. In addition, many research journals, annual conferences, and professional societies around the world are devoted to the field of human factors.

Additional information on the historical developments in human factors is available in Refs. [7–10].

1.2.2   HUMAN RELIABILITY AND ERROR

The history of human reliability and error can be traced back to the late 1950s when H.L. Williams pointed out that the reliability of the human element must be included in the system reliability prediction; otherwise the predicted system reliability would not depict the real picture [11]. In 1960, Shapero et al. pointed out that a large proportion of equipment failures (20%–50%) are due to human error [12]. In the same year, a study conducted by W. I. LeVan also pointed out that 23%–45% of equipment failures were due to human error [13].

In 1973, a well-known journal titled *IEEE Transactions on Reliability* published a special issue on human reliability. In 1986, the first book on human reliability, titled *Human Reliability: With Human Factors,* was published [14]. Additional information on historical developments in human reliability and error is available in Refs. [14–16] and a comprehensive list of publications on the subject up to 1994 is given in Ref. [17].

1.2.3   ENGINEERING MAINTENANCE

Although the history of engineering maintenance may be traced back to the development of the steam engine by James Watt (1735–1819) in 1769 in Great Britain, in the United States a magazine titled *Factory* that first appeared in 1882 played a critical role in the development of the maintenance field [18, 19]. In 1886, a book titled *Maintenance of Railways* was published in the United States [20].

The term *preventive maintenance* was coined in the 1950s and a handbook on maintenance engineering was published in 1957 [21]. Over the years a vast amount of published literature on engineering maintenance in the form of textbooks, technical reports, and articles have appeared and today many institutions throughout the world offer academic programs on engineering maintenance.

1.3   HUMAN RELIABILITY, ERROR, AND HUMAN FACTORS IN ENGINEERING MAINTENANCE-RELATED FACTS AND FIGURES

Some of the facts and figures directly or indirectly concerned with human reliability, error, and human factors in engineering maintenance are as follows:

- Each year U.S. industry spends over $300 billion on plant maintenance and operations, and about 80% of this amount is spent to rectify the chronic failure of people, systems, and machines [1].

- The typical size of a plant maintenance department in manufacturing organizations varied from around 5% to 10% of the total operating workforce [22].
- For the period 1982–1991, a study of safety issues concerning onboard fatalities of jet fleets worldwide reported that maintenance and inspection were the second most important safety issue, with 1481 onboard fatalities [23, 24].
- In 1993, a study of 122 maintenance events involving human factors reported that there were four types of maintenance errors: omissions (56%), wrong installations (30%), wrong parts (8%), and others (6%) [23, 25].
- A study of 213 maintenance problem-related reports, reported that approximately 25% were due to human error [26].
- In 1990, 10 people were killed on the USS *Iwo Jima* (LPH2) naval ship because of a steam leak in the fire room, after maintenance workers repaired a valve and replaced bonnet fasteners with mismatched and wrong material [27].
- A study of maintenance errors in missile operations grouped the error causes under six categories: dials and controls (misread, misset) (38%), wrong installation (28%), loose nuts/fittings (14%), inaccessibility (8%), and miscellaneous (17%) [5, 14].
- In 1985, 520 people were killed in a Japan Airlines Boeing 747 jet accident because of wrong repair [28, 29].
- A study of various maintenance tasks, including removing, adjusting, and aligning, reported an average human reliability of 0.9871 [30].
- A study of 126 human error-related significant events in 1990, in nuclear power generation, reported that 42% of the problems were linked to maintenance and modification [4].
- In 1979, 272 people were killed in a DC-10 aircraft accident in Chicago because of wrong procedures followed by maintenance workers [5].
- A study of over 4400 maintenance history records covering the period from 1992–1994, concerning a boiling water reactor (BWR) nuclear power plant, reported that around 7.5% of all failure records could be classified as human errors related to maintenance actions [31, 32].
- According to Ref. [33] maintenance error contributes to 15% of air carrier accidents and costs the United States industry over 1 billion dollars annually.
- In 1988 in the United Kingdom, 30 people died and 69 were injured seriously at the Clapham Junction railways accident due to a maintenance error in wiring [34].
- A study reported that over 20% of all system failures in fossil power plants occur due to human errors and maintenance errors account for about 60% of the annual power loss due to human errors [35].
- According to a Boeing study, 19.1% of in-flight engine shutdowns are caused by maintenance error [33].
- A study of 199 human errors that occurred in Japanese nuclear power plants from 1965 to 1995 revealed that around 50% of them were related to maintenance activities [36].
- A study reported that maintenance and inspection are the factors in approximately 12% of major aircraft accidents [37, 38].

- In 1988, the upper cabin structure of a Boeing 737-200 aircraft was ripped away during a flight because of structural failure, basically due to the failure of maintenance inspectors to identify over 240 cracks in the aircraft skin during the inspection process [39, 40].

1.4   TERMS AND DEFINITIONS

This section presents some useful terms and definitions directly or indirectly related to human reliability, error, and human factors in engineering maintenance [41–50]:

- **Maintenance.** This is all actions necessary for retaining an item/equipment in, or restoring it to, a specified condition.
- **Human reliability.** This is the probability of accomplishing a specified task successfully by humans at any required stage in system operation within a defined minimum time limit (if the time requirement is specified).
- **Human factors.** This is a body of scientific-related facts concerning the characteristics of humans. The term includes all psychosocial and biomedical considerations. It also includes, but is in no way restricted to, personnel selection, training principles and applications in the area of human engineering, human performance evaluation, aids for task performance, and life support.
- **Human error.** This is the failure to perform a specified task (or the performance of a forbidden action) that could result in disruption of scheduled operations or damage to equipment and property.
- **Corrective maintenance.** This is the unscheduled maintenance or repair to return equipment/system/items to a specified state and performed because maintenance personnel or users perceived deficiencies or failures.
- **Inspection.** This is the qualitative observation of an item's condition or performance.
- **Safety.** This is conservation of human life and its effectiveness, and the prevention of damage to items as per specified mission requirements.
- **Human performance.** This is a measure of actions and failures under given conditions.
- **Preventive maintenance.** This is all actions performed on a planned, periodic, and specific schedule for keeping a piece of equipment in stated working condition through the process of reconditioning and checking. These actions are precautionary steps undertaken for reducing or forestalling the probability of failures or an unacceptable level of degradation in later service, rather than rectifying them after their occurrence.
- **Failure.** This is the inability of an item/equipment/system to perform its stated function.
- **Accident.** This is an event that involves damage to a specified system that suddenly disrupts the current or potential system output.
- **Human performance reliability.** This is the probability that a human will satisfy all stated human functions subject to specified conditions.
- **Maintenance person.** This is an individual who performs preventive maintenance and responds to a user's service call to a repair facility, and

carry out appropriate corrective maintenance on an item/equipment. Some of the other names used for this individual are field engineer, service person, repair person, technician, and mechanic.

- **Continuous task.** This is a task/job that involves some kind of tracking activity (e.g., monitoring a changing situation).
- **Mission time.** This is that element of uptime required to carry out a given mission profile.
- **Hazardous condition.** This is a condition with a potential to threaten human life, properties, health, or the environment.
- **Risk.** This is the probable rate of occurrence of a hazardous condition and the degree of severity of the harm.
- **Maintainability.** This is the probability that a failed item will be restored to satisfactorily working condition.
- **Reliability.** This is the probability that an item will perform its specified function adequately for the desired period when used according to the stated conditions.
- **Redundancy.** This is the existence of more than one means to carry out a stated function.

1.5 USEFUL INFORMATION ON HUMAN RELIABILITY, ERROR, AND HUMAN FACTORS IN ENGINEERING MAINTENANCE

This section lists selected publications, organizations, and data sources that are considered directly or indirectly useful for obtaining information on human reliability, error, and human factors in engineering maintenance.

1.5.1 PUBLICATIONS

These are listed under four distinct classifications: books, technical reports, conference proceedings, and journals.

1.5.1.1 Books

- Reason, J., Hobbs, A., *Managing Maintenance Error: A Practical Guide,* Ashgate Publishing, Aldershot, UK, 2003.
- Dhillon, B.S., *Human Reliability: With Human Factors,* Pergamon Press, New York, 1986.
- Patankar, M.S., Taylor, J.C., *Risk Management and Error Reduction in Aviation Maintenance,* Ashgate Publishing, Aldershot, UK, 2006.
- Whittingham, R.B., *The Blame Machine: Why Human Error Causes Accidents,* Elsevier Butterworth-Heinemann, Oxford, UK, 2004.
- Strauch, B. *Investigating Human Error: Incidents, Accidents, and Complex Systems,* Ashgate Publishing, Aldershot, UK, 2002.
- Corlett, E.N., Clark, T.S., *The Ergonomics of Workspaces and Machines,* Taylor and Francis, London, 1995.
- Karwowski, W., Marras, W.S., *The Occupational Ergonomics Handbook,* CRC Press, Boca Raton, FL, 1999.

- Sanders, M.S., McCormick, E.J., *Human Factors in Engineering and Design,* McGraw Hill, New York, 1993.
- Hall, S., *Railway Accidents,* Ian Allan Publishing, Shepperton, UK, 1997.
- Dhillon, B.S., *Engineering Maintenance: A Modern Approach,* CRC Press, Boca Raton, FL, 2002.

1.5.1.2   Technical Reports

- Report No. CAP 718, Human Factors in Aircraft Maintenance and Inspection, Prepared by the Safety Regulation Group, Civil Aviation Authority, London, UK. Available from the Stationery Office, P.O. Box 29, Norwich, UK.
- Circular 243–AN 151, Human Factors in Aircraft Maintenance and Inspection, International Civil Aviation Organization, Montreal, Canada, 1995.
- Report No. DOT/FRA/RRS-22, Federal Railroad Administration (FRA) Guide for Preparing Accident/Incident Reports, FRA Office of Safety, Washington, D.C., 2003.
- Maintenance Error Decision Aid (MEDA), Developed by Boeing Commercial Airplane Group, Seattle, Washington, 1994.
- Report No. NTSR/SIR-94/02, Maintenance Anomaly Resulting in Dragged Engine During Landing Rollout, Northwest Airlines Flight 18, New Tokyo International Airport, March 2, 1994, National Transportation Safety Board (NTSB), Washington, D.C., 1995.
- Hobbs, A., Williamson, A., Aircraft Maintenance Safety Survey-Results, Report, Australian Transport Safety Bureau, Canberra, Australia, 2000.
- Seminara, J.L., Parsons, S.O., Human Factors Review of Power Plant Maintenance, Report No. EPRI NP-1567, Electric Power Research Institute (EPRI), Palo Alto, CA, 1981.
- WASH-1400, Reactor Safety Study: An Assessment of Accident Risks in U.S. Commercial Nuclear Power Plants, U.S. Nuclear Regulatory Commission, Washington, D.C., 1975.
- Nuclear Power Plant Operating Experience, from the IAEA/NEA Incident Reporting System 1996–1999, Report, Organization for Economic Co-operation and Development (OECD), 2 rue Andre-Pascal, 75775 Paris Cedex 16, France, 2000.
- Report No. DOC 9824-AN/450, Human Factors Guidelines for Aircraft Maintenance Manual, International Civil Aviation Organization (ICAO), Montreal, Canada, 1993.
- Report No. 2–97, Human Factors in Airline Maintenance: A Study of Incident Reports, Bureau of Air Safety Investigation (BASI), Department of Transport and Regional Development, Canberra, Australia, 1997.

1.5.1.3   Conference Proceedings

- Proceedings of the Human Factors and Ergonomics Society Conference, 1997.

- Proceedings of the Airframe/Engine Maintenance and Repair Conference, 1998.
- Proceedings of the Annual Reliability and Maintainability Symposium, 2001.
- Proceedings of the International Conference on Design and Safety of Advanced Nuclear Power Plants, 1992.
- Proceedings of the IEEE 6th Annual Human Factors Meeting, 1997.
- Proceedings of the 5th Federal Aviation Administration (FAA) Meeting on Human Factors Issues in Aircraft Maintenance and Inspection, 1991.
- Proceedings of the 48th Annual International Air Safety Seminar, 1995.
- Proceedings of the 9th International Symposium on Aviation Psychology, 1997.
- Proceedings of the 15th Symposium on Human Factors in Aviation Maintenance, 2001.
- Proceedings of the IEEE International Conference on Systems, Man, and Cybernetics, 1996.

1.5.1.4 Journals

- *International Journal of Industrial Ergonomics*
- *Reliability Engineering and System Safety*
- *Safety Science*
- *ATEC Journal*
- *Human Factors*
- *Rail International*
- *Human Factors in Aerospace and Safety*
- *Maintenance Technology*
- *Industrial Maintenance and Plant Operation*
- *Journal of Quality in Maintenance Engineering*
- *Maintenance Journal*
- *Journal of Occupational Accidents*
- *Aeronautical Journal*
- *International Journal of Man-Machine Studies*
- *Asia Pacific Air Safety*
- *Ergonomics*
- *Aviation Mechanics Bulletin*
- *The CRM Advocate*
- *Applied Ergonomics*
- *Accident Prevention and Analysis*
- *Journal of Railway and Transport*
- *Human Factors and Ergonomics in Manufacturing*
- *Modern Railways*
- *Naval Engineers Journal*
- *Maintenance and Asset Management Journal*
- *Nuclear Safety*

1.5.2 DATA SOURCES

Some of the sources that could be useful, directly or indirectly, in obtaining human reliability, error, and human factors in engineering maintenance-related data are as follows:

- National Technical Information Service, 5285 Port Royal Road, Springfield, Virginia, USA.
- Stewart, C., The Probability of Human Error in Selected Nuclear Maintenance Tasks, Report No. EGG-SSDC-5586, Idaho National Engineering Laboratory, Idaho Falls, Idaho, 1981.
- Gertman, D.I., Blackman, H.S., *Human Reliability and Safety Analysis Data Handbook,* John Wiley and Sons, New York, 1994.
- Data on Equipment Used in Electric Power Generation, Equipment Reliability Information System (ERIS), Canadian Electrical Association, Montreal, Quebec, Canada.
- GIDEP Data, Government Industry Data Exchange Program (GIDEP) Operations Center, Fleet Missile Systems, Analysis, and Evaluation, Department of Navy, Corona, California.
- Schmidtke, H., Editor, *Ergonomic Data for Equipment Design,* Plenum Press, New York, 1984.
- Dhillon, B.S., *Human Reliability: With Human Factors,* Pergamon Press, New York, 1986 (this book lists over 20 sources for obtaining human reliability/error data).
- Boff, K.R., Lincoln, J.E., *Engineering Data Compendium: Human Perception and Performance,* Vols. 1–3, Armstrong Aerospace Medical Research Laboratory, Wright-Patterson Air Force Base, Ohio, 1988.
- Defense Technical Information Center, DTIC-FDAC, 8725 John K. Kingman Road, Suite 0944, Fort Belvoir, Virginia.
- Dhillon, B.S., *Human Error Data Banks, Microelectronics and Reliability,* Vol. 30, 1990, pp. 963–971.
- DOD-HDBK-743A, Anthropometry of U.S. Military Personnel, Department of Defense, Washington, D.C.
- MIL-HDBK-759B, Human Factors Engineering Design for Army Material, Department of Defense, Washington, D.C.

1.5.3 ORGANIZATIONS

Some of the organizations that could be useful, directly or indirectly, to obtain human reliability, error, and human factors in engineering maintenance-related information are as follows:

- International Civil Aviation Organization, 999 University Street, Montreal, Quebec, Canada
- Society for Maintenance and Reliability Professionals, 401 N. Michigan Avenue, Chicago, Illinois.

- Japan Institute of Plant Maintenance, Shuwa Shiba-Koen-3-Chome Bldg., 3-1-38, Shiba-Koen, Minato-Ku, Tokyo, Japan
- Civil Aviation Safety Authority, North Bourne Avenue and Barry Drive Intersection, Canberra, Australia
- Transportation Safety Board of Canada, 330 Spark Street, Ottawa, Ontario, Canada
- Maintenance Engineering Society of Australia (MESA), 11 National Circuit, Barton, ACT, Australia
- Airplane Safety Engineering Department, Boeing Commercial Airline Group, The Boeing Company, 7755 E. Marginal Way South, Seattle, Washington.
- Federal Railroad Administration, 4601 N. Fairfax Drive, Suite 1100, Arlington, Virginia.
- National Research Council, 2101 Second Street, SW, Washington, D.C.
- Society for Machinery Failure Prevention Technology, 4193 Sudley Road, Haymarket, Virginia.
- Transportation Research Board, 2101 Constitution Avenue, NW, Washington, D.C.
- American Institute of Plant Engineers, 539 S. Lexington Place, Anaheim, California.
- Society of Logistics Engineers, 8100 Professional Place, Suite 211, Hyattsville, Maryland.

1.6 SCOPE OF THE BOOK

Just like any other areas of engineering, engineering maintenance is also subjected to human errors. In recent years increasing attention has been given to human errors in the maintenance activity due to various factors, including cost and serious consequences such as the Three Mile Island Nuclear accident and the crash of a DC-10 aircraft at O'Hare Airport in Chicago.

Over the years, a large number of publications on human error, reliability, and human factors in engineering maintenance have appeared basically in the form of journal and conference proceedings articles or technical reports. At present, to the best of the author's knowledge, there is no book that covers all these three topics including maintenance safety within its framework. This book not only attempts to provide up-to-date coverage of the on-going efforts in human reliability, error, and human factors in engineering maintenance, but also covers useful developments in the general areas of human factors, reliability, and error.

Finally, the main objective of this book is to provide professionals concerned with human reliability, error, and human factors in engineering maintenance information that could be useful to minimize or eliminate the occurrence of human error in this area. The book will be useful to many individuals, including engineering professionals working in the area of maintenance, maintenance engineering researchers and instructors, reliability, safety, and human factors professionals, and maintenance engineering administrators.

1.7   PROBLEMS

1. Write an essay on human reliability, error, and human factors in engineering maintenance.
2. Define the following terms:
   - Human factors
   - Human reliability
   - Maintenance
3. List at least five facts and figures concerned with human error/reliability in engineering maintenance.
4. Discuss historical developments in the following two areas, separately:
   - Human reliability
   - Human factors
5. What is the difference between accident and risk?
6. Define the following terms:
   - Human error
   - Maintenance person
7. List at least five journals considered useful for obtaining human reliability and error in engineering maintenance-related information.
8. List at least seven of the most important organizations for obtaining human reliability, error, and human factors in engineering maintenance-related information.
9. What is the difference between preventive and corrective maintenance?
10. List at least six books considered most useful for obtaining, directly or indirectly, human reliability, error, and human factors in engineering maintenance-related information.

REFERENCES

1. Latino, C.J., Hidden Treasure: Eliminating Chronic Failures Can Cut Maintenance Costs up to 60%, Report, Reliability Center, Hopewell, Virginia, 1999.
2. Wu, T.M., Hwang, S.L., Maintenance Error Reduction Strategies in Nuclear Power Plants, Using Root Cause Analysis, *Applied Ergonomics*, Vol. 20, No. 2, 1989, pp. 115–121.
3. Speaker, D.M., Voska, K.J., Luckas, W.J., Identification and Analysis of Human Errors Underlying Electric/Electronic Component Related Events, Report No. NUREG/CR-2987, Nuclear Power Plant Operations, United States Nuclear Regulatory Commission, Washington, D.C., 1983.
4. Reason, J., Human Factors in Nuclear Power Generation: A System's Perspective, *Nuclear Europe Worldscan*, Vol. 17, No. 5–6, 1997, pp. 35–36.
5. Christensen, J.M., Howard, J.M., Field Experience in Maintenance, in *Human Detection and Diagnosis of System Failures*, edited by J. Rasmussen and W.B., Rouse, Plenum Press, New York, 1981, pp. 111–133.
6. Chapanis, A., *Man-Machine Engineering*, Wadsworth Publishing Company, Belmont, California, 1965.
7. Meister, D., Rabideau, G.F., *Human Factors Evaluation in System Development*, John Wiley and Sons, New York, 1965.

8. Woodson, W.E., *Human Factors Design Handbook*, McGraw-Hill Book Company, New York, 1981.
9. McCormick, E.J., Sanders, M.S., *Human Factors in Engineering and Design*, McGraw-Hill Book Company, New York, 1982.
10. Dhillon, B.S., *Advanced Design Concepts for Engineers*, Technomic Publishing Company, Lancaster, Pennsylvania, 1998.
11. Williams, H.L., Reliability Evaluation of the Human Component in Man-Machine Systems, *Electrical Manufacturing*, April 1958, pp. 78–82.
12. Shapero, A., Cooper, J.I., Rappaport, M., Shaeffer, K.H., Bates, C.J., Human Engineering Testing and Malfunction Data Collection in Weapon System Program, WADD Technical Report No. 60–36, Wright-Patterson Air Force Base, Dayton, Ohio, February 1960.
13. LeVan, W.I., Analysis of the Human Error Problem in the Field, Report No. 7-60-932004, Bell Aero Systems Company, Buffalo, New York, June 1960.
14. Dhillon, B.S., *Human Reliability: With Human Factors*, Pergamon Press, New York, 1986.
15. Dhillon, B.S., *Human Reliability and Error in Medical Systems*, World Scientific Publishing, New York, 2003.
16. Dhillon, B.S., *Human Reliability and Error in Transportation Systems*, Springer, London, 2007.
17. Dhillon, B.S., Yang, N., Human Reliability: A Literature Survey and Review, *Microelectronics and Reliability*, Vol. 34, 1994, pp. 803–810.
18. *The Volume Library: A Modern Authoritative Reference for Home and School Use*, The South-Western Company, Nashville, TN, 1993.
19. Factory, McGraw-Hill, New York, 1882–1968.
20. Kirkman, M.M., *Maintenance of Railways*, C.N. Trivess Printers, Chicago, 1886.
21. Morrow, L.C., Editor, *Maintenance Engineering Handbook*, McGraw-Hill, New York, 1994.
22. Niebel, B.W., *Engineering Maintenance Management*, Marcel Dekker, New York, 1994.
23. Human Factors in Airline Maintenance: A Study of Incident Reports, Bureau of Air Safety Inspection, Department of Transport and Regional Development, Canberra, Australia, 1997.
24. Russell, P.D., Management Strategies for Accident Prevention, *Air Asia*, Vol. 6, 1994, pp. 31–41.
25. Circular 243-AN/151, Human Factors in Aircraft Maintenance and Inspection, International Civil Aviation Organization, Montreal, Canada, 1995.
26. Robinson, J.E., Deutsch, W.E., Rogers, J.G., The Field Maintenance Interface between Human Engineering and Maintainability Engineering, *Human Factors*, Vol. 12, 1970, pp. 253–259.
27. *Joint Fleet Maintenance Manual, Vol. 5, Quality Assurance*, Submarine Maintenance Engineering, United States Navy, Portsmouth, New Hampshire, 1991.
28. Gero, D., *Aviation Disasters*, Patrick Stephens, Sparkford, UK, 1993.
29. ASTB Survey of Licensed Aircraft Maintenance Engineers in Australia, Report No. ISBN 0642274738, Australian Transport Safety Bureau (ATSB), Department of Transport and Regional Services, Canberra, Australia, 2001.
30. Sauer, D., Campbell, W.B., Potter, M.R., Askern, W.B., Relationships between Human Resource Factors and Performance on Nuclear Missile Handling Tasks, Report No. AFHRL-TR-76-85/AFWL-TR-76-301, Air Force Human Resources Laboratory/Air Force Weapons Laboratory, Wright-Patterson Air Force Base, Dayton, Ohio, 1976.
31. Pyy, P., An Analysis of Maintenance Failures at a Nuclear Power Plant, *Reliability Engineering and System Safety*, Vol. 72, 2001, pp. 293–302.

32. Pyy, P., Laakso, K., Reiman, L., A Study of Human Errors Related to NPP Maintenance Activities, *Proceedings of the IEEE 6th Annual Human Factors Meeting*, 1997, pp. 12.23–12.28.

33. Marx, D.A., *Learning from Our Mistakes: A Review of Maintenance Error Investigation and Analysis Systems (with Recommendations to the FAA)*, Federal Aviation Administration (FAA), Washington, D.C., January 1998.

34. Report: Investigation into the Clapham Junction Railway Accident, Her Majesty's Stationery Office, London, UK, 1989.

35. Daniels, R.W., The Formula for Improved Plant Maintainability Must Include Human Factors, *Proceedings of the IEEE Conference on Human Factors and Nuclear Safety*, 1985, pp. 242–244.

36. Hasegawa, T., Kemeda, A., Analysis and Evaluation of Human Error Events in Nuclear Power Plants, Presented at the Meeting of the IAEA's CRP on "Collection and Classification of Human Reliability Data for Use in Probabilistic Safety Assessments," May 1998. Available from the Institute of Human Factors, Nuclear Power Engineering Corporation, 3–17-1, Toranomon, Minato-Ku, Tokyo, Japan.

37. Marx, D.A., Graeber, R.C., Human Error in Maintenance, in *Aviation Psychology in Practice*, edited by N. Jonston, N. McDonald, and R. Fuller, Ashgate Publishing, London, 1994, pp. 87–104.

38. Gray, N., Maintenance Error Management in the ADF, *Touchdown* (Royal Australian Navy), December 2004, pp. 1–4. Also available online at http://www.navy.gov.au/publications/touchdown/dec.04/maintrr.html.

39. Report No. DOC 9824-AN450, Human Factors Guidelines for Aircraft Maintenance Manual, International Civil Aviation Organization (ICAO), Montreal, Canada, 2003.

40. Wenner, C.A., Drury, C.G., Analyzing Human Error in Aircraft Ground Damage Incidents, *International Journal of Industrial Ergonomics*, Vol. 26, 2000, pp. 177–1999.

41. Omdahl, T.P., Editor, *Reliability, Availability, and Maintainability (RAM) Dictionary*, ASQC Quality Press, Milwaukee, 1988.

42. AMCP 706-132, *Engineering Design Handbook: Maintenance Engineering Techniques*, Department of Army, Washington, D.C., 1975.

43. DOD INST. 4151.12, Policies Governing Maintenance Engineering within the Department of Defense, Department of Defense, Washington, D.C., June 1968.

44. McKenna, T., Oliverson, R., *Glossary of Reliability and Maintenance Terms*, Gulf Publishing Company, Houston, Texas, 1997.

45. MIL-STD-721C, Definitions of Terms for Reliability and Maintainability, Department of Defense, Washington, D.C.

46. Naresky, J.J., Reliability Definitions, *IEEE Transactions on Reliability*, Vol. 19, 1970, pp. 198–200.

47. Von Alven, W.H., Editor, *Reliability Engineering*, Prentice Hall, Englewood Cliffs, New Jersey, 1964.

48. Meister, D., Human Factors in Reliability, in *Reliability Handbook*, edited by W.G. Ireson, McGraw-Hill, New York, 1966, pp. 12.2–12.37.

49. MIL-STD-721B, Definitions of Effectiveness Terms for Reliability, Maintainability, Human Factors, and Safety, Department of Defense, Washington, D.C., August 1966. Available from the Naval Publications and Forms Center, 5801 Tabor Avenue, Philadelphia, Pennsylvania.

50. MIL-STD-1908, Definitions of Human Factors Terms, Department of Defense, Washington, D.C.

2 Basic Mathematical Concepts

2.1 INTRODUCTION

Although the origin of the word *mathematics* can be traced back to the ancient Greek word *mathema*, which essentially means "science, knowledge, or learning," the history of our current number symbols goes back to around 250 BC to the stone columns erected by the Scythian emperor Asoka of India [1]. Over the centuries, mathematics has developed into various specialized areas, including probability and statistics, applied mathematics, and pure mathematics and is successfully being applied to solve various types of science- and engineering-related problems.

The application of the mathematical concepts in science and engineering ranges from solving interplanetary problems to designing and maintaining engineering equipment in the industrial sector. More specifically, over the past few decades various mathematical concepts, particularly probability distributions and stochastic processes (i.e., Markov modeling), have been used to study various types of problems concerning human reliability and error.

For example, in the late 1960s and early 1970s various statistical distributions were used to represent times to human error [2–4]. Furthermore, in the early 1980s the Markov method was employed to perform human reliability-related analysis of redundant systems [5–7].

This chapter presents various introductory mathematical concepts considered useful for conducting human reliability and error analysis in engineering maintenance.

2.2 BOOLEAN ALGEBRA LAWS AND PROBABILITY PROPERTIES

Boolean algebra is named after an English mathematician, George Boole (1813–1864), who developed it in 1854 [8, 9]. As Boolean algebra plays an important role in human reliability and error-related studies, five of its laws are as follows [10–11]:

ASSOCIATIVE LAW:

$$(A+B)+C = A+(B+C) \qquad (2.1)$$

where A is an arbitrary event or set, B is an arbitrary event or set, C is an arbitrary event or set, and + denotes the union of sets.

$$(A \bullet B) \bullet C = A \bullet (B \bullet C) \qquad (2.2)$$

where the dot (•) denotes the intersections of sets. When Equation (2.2) is written without the dot, it still conveys the same meaning.

COMMUTATIVE LAW:

$$A + B = B + A \tag{2.3}$$

$$AB = BA \tag{2.4}$$

DISTRIBUTIVE LAW:

$$A(B+C) = AB + AC \tag{2.5}$$

$$(A+B)(A+C) = A + BC \tag{2.6}$$

IDEMPOTENT LAW:

$$A + A = A \tag{2.7}$$

$$AA = A \tag{2.8}$$

ABSORPTION LAW:

$$A(A+B) = A \tag{2.9}$$

$$A + (AB) = A \tag{2.10}$$

Probability is the study of random or nondeterministic experiments and it had its real beginnings in the early part of the seventeenth century as a result of investigations of various games of chance by people such as Pierre Fermat (1601–1665) and Blaise Pascal (1623–1662) [12]. The basic properties of probability are presented below [12–15].

• The probability of occurrence of event, say X, is

$$O \leq P(X) \leq 1 \tag{2.11}$$

• The probability of the sample space S is

$$P(S) = 1 \tag{2.12}$$

• The probability of the negation of the sample space S (i.e., $\bar{S}$) is

$$P(\bar{S}) = 0 \tag{2.13}$$

- The probability of occurrence and nonoccurrence of an event, say X, is

$$P(X) + P(\bar{X}) = 1 \qquad (2.14)$$

where $P(X)$ is the occurrence probability of event X and $P(\bar{X})$ is the nonoccurrence probability of event X.

- The probability of an intersection of K independent events is

$$P(Y_1 Y_2 Y_3 \ldots Y_K) = P(Y_1) P(Y_2) P(Y_3) \ldots P(Y_K) \qquad (2.15)$$

where Y_j is the jth event for $j = 1, 2, 3, \ldots, K$ and $P(Y_j)$ is the occurrence probability of event Y_j for $j = 1, 2, 3, \ldots, K$.

- The probability of the union of K independent events is

$$P(X_1 + X_2 + \cdots + X_K) = 1 - \prod_{j=1}^{K} (1 - P(X_j)) \qquad (2.16)$$

where X_j is the jth event; for $j = 1, 2, \ldots, K$ and $P(X_j)$ is the occurrence probability of event X_j; for $j = 1, 2, \ldots, K$.

For $K = 2$, Equation (2.16) reduces to

$$P(X_1 + X_2) = P(X_1) + P(X_2) - P(X_1) P(X_2) \qquad (2.17)$$

- The probability of the union of K mutually exclusive events is

$$P(X_1 + X_2 + \cdots + X_K) = \sum_{j=1}^{K} P(X_j) \qquad (2.18)$$

EXAMPLE 2.1

A maintenance worker is performing a maintenance task composed of two independent steps: X and Y. The task will be accomplished correctly only if both the steps are performed correctly. The probabilities of performing steps X and Y correctly by the maintenance worker are 0.9 and 0.8, respectively. Calculate the probability of accomplishing the task correctly by the maintenance worker.

By substituting the specified data into Equation (2.15), we get

$$\begin{aligned} P(XY) &= P(X) P(Y) \\ &= (0.9)(0.8) \\ &= 0.72 \end{aligned}$$

where $X = Y_1$ and $Y = Y_2$. Thus, the probability of accomplishing the task correctly by the maintenance worker is 0.72.

EXAMPLE 2.2

In Example 2.1, by using Equations (2.14) and (2.17) calculate the probability that the task will not be accomplished successfully by the maintenance worker.

Thus, by using Equation (2.14) and the specified data values, we obtain

$$P(\bar{X}) = 1 - P(X)$$
$$= 1 - 0.9$$
$$= 0.1$$

and

$$P(\bar{Y}) = 1 - P(Y)$$
$$= 1 - 0.8$$
$$= 0.2$$

where $P(\bar{X})$ is the probability of not accomplishing step X correctly by the maintenance worker and $P(\bar{Y})$ is the probability of not accomplishing step Y correctly by the maintenance worker.

Using Equation (2.17) and the above calculated values, we get

$$P(\bar{X} + \bar{Y}) = P(\bar{X}) + P(\bar{Y}) - P(\bar{X})P(\bar{Y})$$
$$= 0.1 + 0.2 - (0.1)(0.2)$$
$$= 0.28$$

where $\bar{X} = X_1$ and $\bar{Y} = X_2$, and $P(\bar{X} + \bar{Y})$ is the probability of not performing steps X or Y correctly. Thus, the probability that the task will not be accomplished successfully by the maintenance worker is 0.28.

2.3   USEFUL DEFINITIONS

This section presents mathematical definitions that are considered useful for performing human reliability and error analysis in engineering maintenance.

2.3.1   PROBABILITY

This is expressed as [14]

$$P(Y) = \lim_{m \to \infty} \left(\frac{M}{m} \right) \tag{2.19}$$

where $P(Y)$ is the probability of occurrence of event Y and M is the total number of times that Y occurs in the m repeated experiments.

2.3.2   CUMULATIVE DISTRIBUTION FUNCTION TYPE I

For continuous random variables, this is defined by [14]

$$F(t) = \int_{-\infty}^{t} f(x)\,dx \qquad (2.20)$$

where $f(t)$ is the probability density function (in human reliability work it is also known as the human error density function), t is the time-continuous random variable, and $F(t)$ is the cumulative distribution function.

2.3.3   PROBABILITY DENSITY FUNCTION TYPE I

For continuous random variables, using Equation (2.20) this is expressed as follows:

$$\frac{d\,F(t)}{dt} = \frac{d\left[\int_{-\infty}^{t} f(x)\,dx\right]}{dt} \qquad (2.21)$$

$$= f(t)$$

2.3.4   PROBABILITY DENSITY FUNCTION TYPE II

For a single-dimension discrete random variable, say X, the discrete probability density function of the random variable X is represented by $f(x_j)$ if the following conditions apply [12]:

$$f(x_j) \geq 0, \quad \text{for all } x_j \in R_x \text{ (range space)}, \qquad (2.22)$$

and

$$\sum_{\substack{all \\ x_j}} f(x_j) = 1 \qquad (2.23)$$

2.3.5   CUMULATIVE DISTRIBUTION FUNCTION TYPE II

For discrete random variables, the cumulative distribution function is expressed by [12]

$$F(x) = \sum_{x_j \leq x} f(x_j) \qquad (2.24)$$

where $F(x)$ is the cumulative distribution function and its value is always $0 \leq F(x) \leq 1$.

2.3.6 RELIABILITY FUNCTION

For continuous random variables, this is expressed by

$$R(t) = 1 - F(t)$$

$$= 1 - \int_{-\infty}^{t} f(x) \, dx \tag{2.25}$$

where $f(x)$ is the failure/human error density function and $R(t)$ is the reliability function.

2.3.7 HAZARD RATE FUNCTION

It is also known as the time-dependent failure/error rate function and is defined by

$$\lambda(t) = \frac{f(t)}{1 - F(t)}$$

$$= \frac{f(t)}{R(t)} \tag{2.26}$$

where $\lambda(t)$ is the hazard rate function or the time-dependent failure/error rate function.

2.3.8 EXPECTED VALUE TYPE I

The expected value, $E(t)$, of a continuous random variable is expressed by [12, 14]

$$E(t) = \mu = \int_{-\infty}^{\infty} t f(t) \, dt \tag{2.27}$$

where μ is the mean value. It is to be noted that in human reliability work, μ is called *mean time to human error*, and f(t) *human error density function*.

2.3.9 EXPECTED VALUE TYPE II

The expected value, $E(x)$, of a discrete random variable x is defined by [12, 14]

$$E(x) = \sum_{j=1}^{k} x_j f(x_j) \tag{2.28}$$

where k is the number of discrete values of the random variable x.

2.3.10 LAPLACE TRANSFORM

The Laplace transform of the function $f(t)$ is defined by

$$F(s) = \int_{0}^{\infty} f(t) e^{-St} \, dt \tag{2.29}$$

where s is the Laplace transform variable, t is the time variable, and $F(s)$ is the Laplace transform of $f(t)$.

EXAMPLE 2.3

Find the Laplace transform of the following function:

$$f(t) = C \tag{2.30}$$

where C is a constant.

Using the above function in Equation (2.29) yields

$$\begin{aligned} F(s) &= \int_0^\infty C e^{-St} dt \\ &= \frac{C e^{-St}}{-s} \Big|_0^\infty \\ &= \frac{C}{s} \end{aligned} \tag{2.31}$$

EXAMPLE 2.4

Find the Laplace transform of the following function:

$$f(t) = e^{-\alpha t} \tag{2.32}$$

where α is a constant. In human reliability work, it is known as the human error rate.

By substituting Equation (2.32) into Equation (2.29) we get

$$\begin{aligned} F(s) &= \int_0^\infty e^{-\alpha t} e^{-St} dt \\ &= \frac{e^{-(S+\alpha)t}}{-(s+\alpha)} \Big|_0^\infty \\ &= \frac{1}{s+\alpha} \end{aligned} \tag{2.33}$$

Table 2.1 presents Laplace transforms of some commonly occurring functions in human reliability-related analysis [16, 17].

2.3.11 LAPLACE TRANSFORM: FINAL-VALUE THEOREM

If the following limits exist, then the final-value theorem may be expressed as

$$\lim_{t \to \infty} f(t) = \lim_{s \to 0} [sF(s)] \tag{2.34}$$

TABLE 2.1
Laplace Transforms of Some Frequently Occurring Functions in Human Reliability-Related Analysis

No.	F(t)	F(s)
1	C, a constant	C/s
2	t^m, for $m = 0, 1, 2, 3, \ldots$	$m!/s^{m+1}$
3	$e^{-\alpha t}$	$1/(s+\alpha)$
4	$te^{-\alpha t}$	$1/(s+\alpha)^2$
5	$\dfrac{df(t)}{dt}$	$sF(s)-f(0)$
6	$tf(t)$	$-\dfrac{dF(s)}{ds}$
7	$\displaystyle\int_0^t f(t)\,dt$	$F(s)/s$
8	$\alpha f_1(t) + \beta f_2(t)$	$\alpha F_1(s) + \beta F_2(s)$
9	$t^{m-1}/(m-1)!$	$\dfrac{1}{s^m}$, $m = 1, 2, 3, \ldots$

2.4 PROBABILITY DISTRIBUTIONS

In human reliability-related analysis various types of discrete and continuous random variable probability distributions are used. Some examples of these distributions are binomial, Poisson, exponential, and normal distribution. This section presents probability distributions that are considered useful for application in performing human reliability and error analysis in engineering maintenance [18].

2.4.1 POISSON DISTRIBUTION

The Poisson distribution is a discrete random variable distribution and is named after Simeon Poisson (1781–1840) [1]. The distribution is used in situations when one is concerned with the occurrence of a number of events that are of the same kind. The occurrence of an event is denoted as a point on a time scale, and in human reliability work an event denotes a human error. The distribution density function is expressed by

$$f(K) = \frac{(\alpha t)^K e^{-\alpha t}}{K!}, \quad \text{for } K = 0, 1, 2, 3, \ldots \tag{2.35}$$

where t is time and α is the constant arrival or error rate.
    The cumulative distribution function, F, is

$$F = \sum_{j=0}^{K} [(\alpha t)^j e^{-\alpha t}]/j! \tag{2.36}$$

The distribution mean is given by [15, 18]

$$\mu_p = \alpha t \tag{2.37}$$

where μ_p is the mean of the Poisson distribution.

2.4.2 BINOMIAL DISTRIBUTION

This is another discrete random variable distribution. The distribution is also known as the Bernoulli distribution, after Jakob Bernoulli (1654–1705), its originator [1]. The distribution is used in situations when one is interested in the probability of outcome such as the total number of errors/failures in a sequence of, say K, trials. The distribution is based on the condition that each trial has two possible outcomes (e.g., success and failure), and each trial's probability remains constant.

The distribution probability density function, $f(x)$, is defined by

$$f(x) = \binom{K}{j} p^x q^{K-x}, \quad \text{for } x = 0, 1, 2, 3, ..., K. \tag{2.38}$$

where

$$\binom{K}{j} = \frac{K!}{j!(K-j)!}$$

x is the total number of failures/errors in K trials, q is the probability of failure of a single trial, and p is the probability of success of a single trial.

The cumulative distribution function is given by

$$F(x) = \sum_{j=0}^{x} \binom{K}{j} p^j q^{K-j} \tag{2.39}$$

where $F(x)$ is the probability of x or less failures (errors) in K trials.

The distribution mean is given by [18]

$$\mu_b = Kp \tag{2.40}$$

where μ_b is the mean of the binomial distribution.

2.4.3 GEOMETRIC DISTRIBUTION

This discrete random variable distribution is based on the same assumptions as the binomial distribution, except that the number of trials is not fixed. More specifically, all trials are independent and identical and each can result in one of the two possible outcomes (i.e., a success or a failure (error)). The distribution probability density function, $f(x)$, is defined by [13, 19]

$$f(x) = pq^{x-1}, \quad \text{for } x = 1, 2, 3, ... \tag{2.41}$$

The cumulative distribution function is given by

$$F(x)=\begin{cases} 0, & x<1 \\ \displaystyle\sum_{x_j\le[x]} pq^{x_j-1}, & x\ge1 \end{cases} \tag{2.42}$$

The distribution mean is given by

$$\mu_g=\frac{1}{p} \tag{2.43}$$

where μ_g is the mean of the geometric distribution.

2.4.4 EXPONENTIAL DISTRIBUTION

This is probably the most widely used continuous random variable probability distribution in performing reliability studies, because many engineering parts exhibit constant failure rate during their useful life period [20].

The distribution probability density function is defined by

$$f(t)=\lambda e^{-\lambda t} \quad t\ge0, \lambda>0 \tag{2.44}$$

where $f(t)$ is the probability density function (in reliability work, it is also called failure density function or error density function), λ is the distribution parameter (in human reliability work, it is known as the constant human error rate), and t is time.

Using Equations (2.20) and (2.44), we obtain the following expression for the cumulative distribution function:

$$F(t)=\int_0^t \lambda e^{-\lambda t}\, dt$$
$$=1-e^{-\lambda t} \tag{2.45}$$

By substituting Equation (2.44) into Equation (2.27), we get the following expression for the distribution expected or mean value:

$$E(t)=\mu=\int_0^\infty t\lambda e^{-\lambda t}\, dt$$
$$=\frac{1}{\lambda} \tag{2.46}$$

EXAMPLE 2.5

Assume that the constant error rate of maintenance personnel in performing a certain maintenance task is 0.009 errors/hour. Calculate the probability that the maintenance personnel will make an error during an 8-hour mission.

By substituting the specified data values into Equation (2.45), we get

$$F(8) = 1 - e^{-(0.009)(8)}$$
$$= 0.0695$$

Thus, the probability that the maintenance personnel will make an error during the specified time period is 0.0695.

2.4.5 Normal Distribution

This is a widely used continuous random variable probability distribution and sometimes it is also called the Gaussian distribution after Carl Friedrich Gauss (1777–1855), the German mathematician. The probability density function of the distribution is defined by

$$f(t) = \frac{1}{\sigma\sqrt{2\pi}} \exp\left[-\frac{(t-\mu)^2}{2\sigma^2}\right], \quad -\infty \langle t \langle +\infty \tag{2.47}$$

where μ and σ are the distribution parameters (i.e., mean and standard deviation, respectively).

Substituting Equation (2.47) into Equation (2.20), we obtain the following cumulative distribution function:

$$F(t) = \frac{1}{\sigma\sqrt{2\pi}} \int_{-\infty}^{t} \exp\left[-\frac{(x-\mu)^2}{2\sigma^2}\right] dx \tag{2.48}$$

Using Equation (2.47) in Equation (2.27), we obtain the following expression for the distribution expected or mean value:

$$E(t) = \mu \tag{2.49}$$

2.4.6 Gamma Distribution

This is another continuous random variable probability distribution and is quite flexible in fitting a wide range of problems including human errors. The distribution probability density function is defined by

$$f(t) = \frac{\lambda(\lambda t)^{\alpha-1} e^{-\lambda t}}{\Gamma(x)}, \quad t \geq 0, \lambda, \alpha \rangle 0 \tag{2.50}$$

where $\Gamma(\cdot)$ is the gamma function, λ is the distribution scale parameter, and α is the distribution shape parameter.

Using Equation (2.50) in Equation (2.20), we get the following cumulative distribution function:

$$F(t) = 1 - \sum_{j=0}^{\alpha-1} \frac{e^{-\lambda t}(\lambda t)^j}{j!} \tag{2.51}$$

Substituting Equation (2.50) into Equation (2.27), we get the following expression for the distribution expected or mean value:

$$E(t) = \frac{\alpha}{\lambda} \tag{2.52}$$

It is to be noted that at $\alpha = 1$, the gamma distribution becomes the exponential distribution.

2.4.7 RAYLEIGH DISTRIBUTION

This continuous random variable probability distribution is often used in the theory of sound and in reliability studies and is known after John Rayleigh (1842–1919), its originator [1]. The distribution probability density function is defined by

$$f(t) = \frac{2}{\beta^2} t e^{-\left(\frac{t}{\beta}\right)^2}, \quad t \geq 0, \beta \rangle 0 \tag{2.53}$$

where β is the distribution parameter.

Substituting Equation (2.53) into Equation (2.20), we obtain the following cumulative distribution function:

$$F(t) = 1 - e^{-\left(\frac{t}{\beta}\right)^2} \tag{2.54}$$

Using Equation (2.53) in Equation (2.27), we obtain the following equation for the distribution expected or mean value:

$$E(t) = \beta \Gamma\left(\frac{3}{2}\right) \tag{2.55}$$

where $\Gamma(\cdot)$ is the gamma function and is defined by

$$\Gamma(y) = \int_0^{\infty} t^{y-1} e^{-t} dt, \quad \text{for } y \rangle 0 \tag{2.56}$$

2.4.8 WEIBULL DISTRIBUTION

This continuous random variable probability distribution can be used to represent many different physical phenomena and it was developed by W. Weibull, a Swedish mechanical engineering professor, in the early 1950s [21]. The distribution probability density function is expressed by

$$f(t) = \frac{\theta t^{\theta-1}}{\beta^{\theta}} e^{-\left(\frac{t}{\beta}\right)^{\theta}}, \quad t \geq 0, \theta, \beta \rangle 0 \tag{2.57}$$

where θ is the distribution shape parameter and β is the distribution scale parameter.

Using Equations (2.57) and (2.20), we get the following cumulative distribution function:

$$F(t) = 1 - e^{-\left(\frac{t}{\beta}\right)^{\theta}} \qquad (2.58)$$

Substituting Equation (2.57) into Equation (2.27), we get the following expression for the distribution expected or mean value:

$$E(t) = \beta \Gamma \left(1 + \frac{1}{\theta}\right) \qquad (2.59)$$

It is to be noted that for $\theta = 1$ and 2, the exponential and Rayleigh distributions are the special cases of the Weibull distribution, respectively.

2.5 SOLVING FIRST-ORDER DIFFERENTIAL EQUATIONS USING LAPLACE TRANSFORMS

Sometime in human reliability and error studies, the Markov method (described in Chapter 4), is used and it results in a system of linear first-order differential equations. The use of Laplace transforms is considered to be an effective approach to find solutions to these differential equations. The following example demonstrates the application of Laplace transforms in finding the solutions to a set of linear first-order differential equations.

EXAMPLE 2.6

Assume that an engineering system can either be in three states: operating normally, failed due to hardware problems, or failed due to maintenance errors. The following set of differential equations describes the system:

$$\frac{dP_0(t)}{dt} + (\lambda + \lambda_m) P_0(t) = 0 \qquad (2.60)$$

$$\frac{dP_1(t)}{dt} - \lambda P_0(t) = 0 \qquad (2.61)$$

$$\frac{dP_2(t)}{dt} - \lambda_m P_0(t) = 0 \qquad (2.62)$$

At $t = 0$, $P_0(0) = 1$, and $P_1(0) = P_2(0) = 0$, where $P_i(t)$ is the probability that the engineering system is in state i at time for $i = 0$ (operating normally), $i = 1$ (failed due to hardware problems), $i = 2$ (failed due to maintenance errors); λ is the constant failure rate of the system due to hardware problems; and λ_m is the constant failure rate of the system due to maintenance errors.

Find solutions to differential Equations (2.60)–(2.62) by using Laplace transforms.

Thus, taking Laplace transforms of Equations (2.60)–(2.62), using the given initial conditions, and then solving the resulting equations, we get

$$P_0(s) = \frac{1}{(s + \lambda + \lambda_m)} \qquad (2.63)$$

$$P_1(s) = \frac{\lambda}{s(s + \lambda + \lambda_m)} \qquad (2.64)$$

$$P_2(s) = \frac{\lambda_m}{s(s + \lambda + \lambda_m)} \qquad (2.65)$$

where s is the Laplace transform variable, and $P_i(s)$ is the Laplace transform of the probability that the engineering system is in state i at time t, for $i = 0, 1, 2$.

By taking the inverse Laplace transforms of Equations (2.63)–(2.65), we obtain

$$P_0(t) = e^{-(\lambda + \lambda_m)t} \qquad (2.66)$$

$$P_1(t) = \frac{\lambda}{\lambda + \lambda_m} (1 - e^{-(\lambda + \lambda_m)t}) \qquad (2.67)$$

$$P_2(t) = \frac{\lambda_m}{\lambda + \lambda_m} (1 - e^{-(\lambda + \lambda_m)t}) \qquad (2.68)$$

Thus, Equations (2.66)–(2.68) represent solutions to differential Equations (2.60)–(2.62).

2.6 PROBLEMS

1. Prove Equation (2.6).
2. Assume that a maintenance worker is performing a maintenance task composed of three independent steps: steps X, Y, and Z. The task will be accomplished correctly only if all the steps are performed correctly. The probabilities of performing steps X, Y, and Z correctly by the maintenance worker are 0.95, 0.75, and 0.99, respectively. Calculate the probability of accomplishing the task correctly by the maintenance worker.
3. In the above Problem No. 2, by using Equations (2.14) and (2.16) calculate the probability that the task will not be accomplished successfully by the maintenance worker.
4. Define *probability* mathematically.
5. Take the Laplace transform of the following function:

$$f(t) = te^{-\lambda t} \qquad (2.69)$$

where λ is a constant and t is a time variable.

6. Obtain an expression for the hazard rate by using the following failure density function.

$$f(t) = \lambda e^{-\lambda t} \quad t \geq 0, \lambda \rangle 0 \tag{2.70}$$

   where λ is the distribution parameter and t is time.
7. Prove Equation (2.46).
8. What are the special case probability distributions of the Weibull distribution?
9. Assume that the constant error rate of maintenance personnel in performing a certain task is 0.001 errors/hour. Calculate the probability that the maintenance personnel will not make an error during a 6-hour mission.
10. Prove that the sum of Equations (2.62)–(2.65) is equal to $1/S$. Comment on the end result.

REFERENCES

1. Eves, H., *An Introduction to the History of Mathematics*, Holt, Reinhart, and Winston, New York, 1976.
2. Askren, W.B., Regulinski, T.L., Quantifying Human Performance for Reliability Analysis of Systems, *Human Factors*, Vol. 11, 1969, pp. 393–396.
3. Regulinski, T.L., Askren, W.B., Mathematical Modeling of Human Performance Reliability, *Proceedings of the Annual Symposium on Reliability*, 1969, pp. 5–11.
4. Regulinski, T.L., Askren, W.B., Stochastic Modeling of Human Performance Effectiveness Functions, *Proceedings of the Annual Reliability and Maintainability Symposium*, 1972, pp. 407–416.
5. Dhillon, B.S., Stochastic Models for Predicting Human Reliability, *Microelectronics and Reliability*, Vol. 25, 1982, pp. 491–496.
6. Dhillon, B.S., Rayapati, S.N., Reliability Analysis of Non-Maintained Parallel Systems Subject to Hardware Failure and Human Error, *Microelectronics and Reliability*, Vol. 25, 1985, pp. 111–122.
7. Dhillon, B.S., System Reliability Evaluation Models with Human Errors, *IEEE Transactions on Reliability*, Vol. 32, 1983, pp. 47–48.
8. Boole, G., *An Investigation of the Laws of Thought*, Dover Publications, New York, 1951.
9. Hailperin, T., *Boole's Logic and Probability*, North Holland, Amsterdam, 1986.
10. Lipschutz, S., *Set Theory*, McGraw-Hill, New York, 1964.
11. Fault Tree Handbook, Report No. NUREG-0492, U.S. Nuclear Regulatory Commission, Washington, D.C., 1981.
12. Lipschutz, S., *Probability*, McGraw-Hill, New York, 1965.
13. Montgomery, D.C., Runger, G.C., *Applied Statistics and Probability for Engineers*, John Wiley and Sons, New York, 1999.
14. Mann, N.R., Schafer, R.E., Singpurwalla, N.D., *Methods for Statistical Analysis of Reliability and Life Data*, John Wiley and Sons, New York, 1974.
15. Shooman, M.L., *Probabilistic Reliability: An Engineering Approach*, McGraw-Hill, New York, 1968.
16. Spiegel, M.R., *Laplace Transforms*, McGraw-Hill, New York, 1965.

17. Oberhettinger, F., Badii, L., *Tables of Laplace Transforms*, Springer-Verlag, New York, 1973.
18. Patel, J.K., Kapadia, C.H., Owen, D.B., *Handbook of Statistical Distributions*, Marcel Dekker, New York, 1976.
19. Tsokos, C.P., *Probability Distributions: An Introduction to Probability Theory with Applications*, Wadsworth Publishing Company, Belmont, California, 1972.
20. Davis, D.J., An Analysis of Some Failure Data, *Journal of the American Statistical Association*, June 1952, pp. 113–150.
21. Weibull, W., A Statistical Distribution Function of Wide Applicability, *Journal of Applied Mechanics*, Vol. 18, 1951, pp. 293–297.

3 Introductory Human Factors, Reliability, and Error Concepts

3.1 INTRODUCTION

Over the years considerable new developments have taken place in the areas of human factors, reliability, and error. Human factors, reliability, and error have become recognizable disciplines in the industrial sector in many parts of the world. There are many standard documents on human factors that directly or indirectly cover human reliability and error as well. These standard documents are often cited in the design specification of complex engineering systems [1].

More specifically, the new system design must satisfy requirements specified in these documents. Thus, nowadays it is not uncommon to come across human factors specialists (who cover human reliability and error as well) working alongside design engineers during the design and development of engineering systems, for use in areas such as nuclear power generation and aviation. These specialists use various human factors, reliability, and error-related concepts to produce effective systems with respect to humans [2, 3].

This chapter presents various introductory human factors, reliability, and error concepts considered useful for application in the areas of engineering maintenance, taken from published literature.

3.2 HUMAN FACTORS OBJECTIVES AND MAN–MACHINE SYSTEM TYPES AND COMPARISONS

There are many objectives of human factors. They may be categorized under four distinct classifications as follows [4]:

- **Classification I: Fundamental Operational Objectives.** These are basically concerned with improving system performance, increasing safety, and reducing human errors.
- **Classification II: Objectives Affecting Operators and Users.** These are concerned with improving the work environment, increasing aesthetic appearance, increasing user acceptance and ease of use, and reducing fatigue, physical stress, boredom, and monotony.
- **Classification III: Objectives Affecting Reliability and Maintainability.** These are concerned with improving reliability, increasing maintainability, reducing the manpower need, and reducing training requirements.

- **Classification IV: Miscellaneous Objectives.** These are concerned with items such as reducing equipment and time losses and increasing production economy.

Although there are many types of man-machine systems, they may be grouped under the following three categories [5]:

- **Category I: Automated Systems.** These systems carry out operation-related functions including processing, sensing, decision making, and action. The majority of these systems are of the closed-loop type and normally the basic human functions associated with such systems are monitoring, maintenance, and programming.
- **Category II: Mechanical or Semiautomatic Systems.** These systems contain well-integrated parts, such as various types of powered machine tools. Normally, in these systems the machines provide the power and the humans typically carry out the control function.
- **Category III: Manual Systems.** These systems contain hand tools and other aids along with the human operator who controls the overall operation. The operator makes use of his or her own physical energy as a power source, and then transmits/receives from the tools a significant amount of information.

Some of the important comparisons between humans and machines (in parentheses) are as follows [6]:

- Humans have excellent memory (machines are remarkably costly to have the same capability).
- Humans have relatively easy maintenance needs (machines' maintenance problems become serious with the increase in complexity).
- Humans are subjected to social environments of all kinds (machines are independent of social environments of all types).
- Humans' performance efficiency is affected by anxiety (machines are quite independent of this shortcoming).
- Humans are very flexible with respect to task performance (machines are relatively inflexible).
- Humans have high tolerance for factors such as ambiguity, vagueness, and uncertainty (machines are quite limited in tolerance in regard to factors such as these).
- Humans are limited to a certain degree in channel capacity (machines have unlimited channel capacities).
- Humans are poor monitors of events that do not occur frequently (machines possess options to be designed to reliably detect infrequently occurring events.
- Humans are subjected to stress because of interpersonal or other difficulties (machines are completely free of such difficulties).

- Humans are unsuitable for performing tasks such as amplification, data coding, or transformation (machines are extremely useful for performing tasks such as these).
- Humans have rather restricted short-term memory for factual matters (machines can have unlimited short-term memory but its affordability is a limiting factor).
- Humans are subjected to factors such as motion sickness, disorientation, and Coriolis effects (machines are completely free of such effects).
- Humans are often subjected to departure from following an optimum strategy (machines always follow the design strategy).
- Humans are subjected to deterioration in performance because of boredom and fatigue (machines are not affected by factors such as these, but their performance is subjected to deterioration because of wear or lack of calibration).
- Humans are very capable of making inductive decisions under novel conditions (machines possess very little or no induction capabilities at all).

3.3 HUMAN SENSORY CAPACITIES AND TYPICAL HUMAN BEHAVIORS AND THEIR CORRESPONDING DESIGN CONSIDERATIONS

Humans possess many useful sensors: touch, sight, taste, hearing, and smell. A clear understanding of their sensory capacities can be quite useful in reducing the occurrence of human errors in engineering maintenance. Thus, some of the human sensory-related capacities are described below [3, 7].

3.3.1 TOUCH

The sense of touch is related to humans' ability in interpreting visual and auditory stimuli. The sensory cues received by muscles and the skin can be used for sending messages to the brain, thus relieving the ears and the eyes of the workload, to a certain degree.

3.3.2 SIGHT

This is stimulated by the electromagnetic radiation of certain wavelengths, often referred to as the visible portion of the electromagnetic spectrum. The spectrum's various areas, as seen by the human eyes, appear to vary in brightness. For example, in the day light, the human eyes are very sensitive to greenish-yellow light with a wavelength of about 5500 Angstrom units [7].

Moreover, the human eyes perceive all colors when they are looking straight ahead but as the viewing angle increases, the color perception begins to decrease. Also, the human eyes see differently from different angles.

3.3.3　Vibration

Past experiences indicate that the presence of vibration could be quite detrimental to the performance of mental and physical tasks by humans such as maintenance personnel. There are numerous vibration parameters including frequency, velocity, acceleration, and amplitude. More specifically, large amplitude and low frequency vibrations contribute to various problems including headaches, eyestrain, fatigue, motion sickness, and interference with the ability to read and interpret instruments properly [7].

Furthermore, high frequency and low amplitude vibrations can also cause fatigue to a certain degree.

3.3.4　Noise

Noise may simply be described as sounds that lack coherence and human reactions to noise extend beyond the auditory systems (e.g., irritability, fatigue, or boredom). Excessive noise can lead to problems such as adverse effects on tasks requiring a high degree of muscular coordination and precision or intense concentration, reduction in the workers' efficiency, and loss of hearing if exposed for long periods.

Over the years, various human behaviors have been observed by researchers in the field. Some of the typical human behaviors and their corresponding design considerations are presented in Table 3.1 [2].

TABLE 3.1
Typical Human Behaviors and Their Corresponding Design Considerations

No.	Typical Human Behavior	Corresponding Design Consideration
1	Humans often tend to hurry	Develop design such that it properly takes into consideration the element of human hurry
2	Humans get easily confused with unfamiliar items/things	Avoid designing totally unfamiliar items/things
3	Humans often use their sense of touch for exploring or testing the unknown	Give careful attention to this factor during design, particularly to the product/item handling aspect
4	Humans frequently regard manufactured items as being safe	Design products such that they become impossible to be used incorrectly
5	Humans have become accustomed to certain color meanings	During design strictly observe existing color coding standards
6	Humans normally expect to turn on the electrical power, the switches have to move upward, or to the right, etc.	Design such switches as per human expectations
7	Humans always expect that faucets/handles will rotate counter-clockwise for increasing the flow of gas, steam, or liquid	Design such items as per human expectations

3.4   HUMAN FACTORS–RELATED FORMULAS

Over the years, researchers have developed various types of mathematical formulas for estimating human factors–related information. Four of these formulas considered useful for application in engineering maintenance are presented below.

3.4.1   FORMULA FOR ESTIMATING INSPECTOR PERFORMANCE

This formula is concerned with estimating inspector performance with respect to inspection tasks. Thus, the inspector performance is expressed by [3, 8]

$$\theta_i = \frac{T_{tr}}{n_p - n_{ie}} \tag{3.1}$$

where θ_i is the inspector performance expressed in minutes per correct inspection, n_p is the total number of patterns inspected, n_{ie} is the total number of inspector errors, and T_{tr} is the total reaction time expressed in minutes.

3.4.2   FORMULA FOR ESTIMATING REST PERIOD

When humans perform lengthy or strenuous tasks, the incorporation of proper rest periods is considered essential. Thus, this formula is concerned with estimating the length of scheduled or unscheduled rest periods. The length of the required rest period is expressed by [9]

$$T_{rp} = \frac{T_w (E_a - E_s)}{(E_a - RL_a)} \tag{3.2}$$

where T_{tr} is the required length of the rest period expressed in minutes, T_w is the working time expressed in minutes, E_a is the average energy cost/expenditure expressed in kilocalories per minute of work, E_s is the kilocalories per minute adopted as standard, and RL_a is the approximate resting level expressed in kilocalories per minute (usually, the value of RL_a is taken as 1.5).

3.4.3   FORMULA FOR ESTIMATING CHARACTER HEIGHT

As usually the instrument panels are located at a viewing distance of 28 inches for the comfortable performance and control of adjustment-oriented tasks, this formula is concerned with estimating the character height at the viewing distance of 28 inches. Thus, the character height is expressed by

$$C_h = \frac{C_s D_v}{28} \tag{3.3}$$

where D_v is the specified viewing distance expressed in inches, C_h is the character height at the specified viewing distance, D_v, expressed in inches, and C_s is the standard character height from a viewing distance of 28 inches.

EXAMPLE 3.1

Assume that maintenance workers have to read a meter from a distance of 70 inches and the standard character height at a viewing distance of 28 inches is 0.50 inches. Estimate the height of numerals for the stated viewing distance.

By substituting the given data values into Equation (3.3), we get

$$C_h = \frac{(0.50)(70)}{28}$$
$$= 1.25 \text{ inches}$$

Thus, the height of numerals for the stated viewing distance of 70 inches is 1.25 inches.

3.4.4 FORMULA FOR ESTIMATING GLARE CONSTANT

Various types of human errors can occur in maintenance work due to glare. The value of the glare constant can be estimated by using the following formula [9]:

$$\alpha = \frac{(\lambda^{0.8})(\beta^{1.6})}{L_g \mu^2} \tag{3.4}$$

where α is the glare constant, L_g is the general background luminance, λ is the solid angle subtended at the eye by the source, μ is the angle between the direction of the glare source and the viewing direction, and β is the source luminance.

3.5 USEFUL HUMAN FACTORS GUIDELINES AND DATA COLLECTION SOURCES

Over the years, researchers working in the area of human factors have developed many useful human factors-related guidelines for application in engineering system design. Some of these guidelines are as follows [2, 6]:

- Review system objectives with respect to human factors.
- Obtain all appropriate human factors-related design reference documents.
- Develop an effective human factors-related checklist for use during system design and operation phases.
- Use the services of human factors experts as considered appropriate.
- Conduct field tests of the system design prior to its approval for delivery to customers.
- Review final production drawings in regard to human factors.
- Make use of mock-ups for "testing" the effectiveness of user-hardware interface designs.

There are many sources for collecting human factors-related data. Some of the important ones are as follows [10, 11]:

- **Test reports.** These reports contain data obtained from testing manufactured items or goods.
- **User experience reports.** These reports contain data reflecting experiences of users with the system/equipment in the field use environment.
- **Published standards.** These documents are published by various organizations including professional societies and government agencies.
- **Published literature.** This includes items such as journals, technical reports, and conference proceedings.
- **System development phase.** This is a good source for collecting various types of human factors-related data.
- **Previous experience.** This is a quite good source for obtaining data from similar cases that have occurred in the past.

3.6 HUMAN PERFORMANCE EFFECTIVENESS AND OPERATOR STRESS CHARACTERISTICS

Over the years, various researchers have studied the relationship between human performance and stress. They conclude that such relationship basically follows the shape of the curve shown in Figure 3.1 [12, 13].

The curve shows that stress to a moderate level is necessary to achieve optimal human performance effectiveness. Otherwise, at a very low stress, the task will become dull and unchallenging, and consequently human performance effectiveness will not be at its highest point.

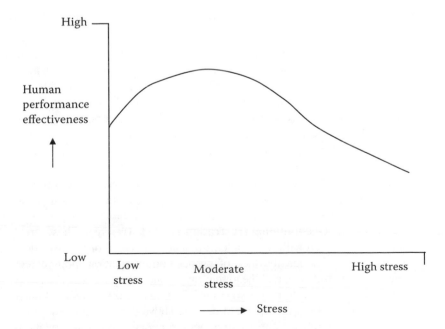

FIGURE 3.1 Human performance effectiveness versus stress curve.

In contrast, stress beyond a moderate level will cause deterioration in human performance because of factors such as fear, worry, or other kinds of psychological stress. It simply means that the probability of human error occurrence will be higher under high stress than under moderate stress.

Human operators perform various types of tasks in diverse engineering areas. In performing such tasks, they may have certain limitations. Past experiences indicate that when these limitations are violated, probability for the error increases quite significantly [14]. This probability can be reduced significantly by carefully considering operator limitations or characteristics during the system design. Some of these characteristics are as follows [14]:

- Performing task steps at high speed
- Poor feedback information in determining the correctness of actions taken
- The requirement for prolonged monitoring
- Having rather short decision-making time
- Performing tasks that require a very long sequence of steps
- Requirement to operate more than one control simultaneously at high speed
- Requirement to make quick comparisons of two or more displays
- Requirement to make decisions on the basis of data collected from diverse sources

3.7 OCCUPATIONAL STRESSORS AND GENERAL STRESS FACTORS

The occupational stressors may be classified under the following four categories [12]:

- **Category I: Workload-related stressors.** These stressors are concerned with work under load or work overload. In the case of work under load, the present duties being carried out by the individual fail to provide sufficient stimulation. Some examples of work under load are the lack of any intellectual input, task repetitiveness, and the lack of opportunity to use acquired expertise and skills of the individual. In contrast, in the case of work overload the job requirements exceed the ability of the individual to satisfy them in an effective manner.
- **Category II: Occupational change-related stressors.** These stressors are concerned with factors that disrupt cognitive, behavioral, and physiological patterns of functioning of the individual.
- **Category III: Occupational frustration-related stressors.** These stressors are concerned with the problems related to occupational frustration. The problems include the lack of proper communication, poor career development guidance, and the ambiguity of one's role.
- **Category IV: Miscellaneous stressors.** These stressors include all other stressors that are not incorporated into the above three categories. Some examples of the miscellaneous stressors are poor interpersonal relationships, too much or too little lighting, and too much noise.

Over the years, various researchers in the area of human engineering have pointed out that there are many general factors that considerably increase stress on an individual, in turn leading to a significant deterioration in his or her reliability. Some of these general factors are as follows [15]:

- Poor health
- Possibility of redundancy at work
- Having to work with individuals with unpredictable temperaments
- Serious financial difficulties
- Working under extremely tight time pressures
- Lacking the proper expertise to perform the ongoing job
- Experiencing difficulties with spouse or children or both
- Poor chances for promotion
- Excessive demands from superiors at work

3.8 HUMAN PERFORMANCE RELIABILITY AND CORRECTABILITY FUNCTIONS

Both these functions are derived below, separately.

3.8.1 HUMAN PERFORMANCE RELIABILITY FUNCTION

Although all the tasks performed by humans are not in continuous time, from time to time humans do perform time-continuous tasks such as scope monitoring, missile countdown, and aircraft maneuvering. In situations such as these, human performance reliability is a very important parameter.

Thus, in time-continuous tasks the probability of occurrence of human error in the finite time interval Δt is expressed by [16–19]

$$P(B/A) = \lambda(t)\Delta t \tag{3.5}$$

where A is an errorless performance event of duration time t, B is an event in which the human error will occur in time interval $(t, t + \Delta t)$, and $\lambda(t)$ is the time-dependent error rate.

Thus, the joint probability of the errorless human performance may be expressed as follows:

$$P(\bar{B}/A)P(A) = P(A) - P(B/A)P(A) \tag{3.6}$$

where $P(A)$ is the probability of occurrence of event A, and $\bar{B}$ is the event that human error will not occur in time interval $[t, t + \Delta t]$.

Equation (3.6) may be rewritten as follows [16–19]:

$$HR(t) - HR(t)P(B/A) = HR(t + \Delta t) \tag{3.7}$$

where $HR(t)$ is the human reliability at time t and $HR(t + \Delta t)$ is the human reliability at time $t + \Delta t$.

It is to be noted that Equation (3.6) denotes an errorless human performance probability over time intervals $[0, t]$ and $[t, t + \Delta t]$.

By substituting Equation (3.5) into Equation (3.7), we obtain

$$\frac{HR(t + \Delta t) - HR(t)}{\Delta t} = -\lambda(t)\, HR(t) \qquad (3.8)$$

In the limiting case, Equation (3.8) becomes

$$\frac{dHR(t)}{dt} = -\lambda(t)\, HR(t) \qquad (3.9)$$

By rearranging Equation (3.9), we get

$$\frac{1}{HR(t)}\, dHR(t) = -\lambda(t)\, dt \qquad (3.10)$$

By integrating both sides of Equation (3.10) over the time interval $[0, t]$, we obtain

$$\int_{1}^{HR(t)} \frac{1}{HR(t)} \cdot dHR(t) = -\int_{0}^{t} \lambda(t)\, dt \qquad (3.11)$$

because at $t = 0$, $HR(0) = 1$.

After evaluating the left-hand side of Equation (3.11), we get

$$\ln HR(t) = -\int_{0}^{t} \lambda(t)\, dt \qquad (3.12)$$

Thus, from Equation (3.12), we obtain

$$HR(t) = e^{-\int_{0}^{t} \lambda(t)\, dt} \qquad (3.13)$$

Equation (3.13) is the general expression for computing human reliability, irrespective of whether the human error rate is constant or nonconstant. More specifically, it holds when time to human error is described by statistical distributions such as normal, gamma, exponential, Weibull, and Rayleigh.

EXAMPLE 3.2

Assume that the time to human error of a maintenance worker follows Weibull distribution. Thus, his or her time-dependent error rate is expressed by

$$\lambda(t) = \frac{\beta t^{\beta - 1}}{\theta^{\beta}} \qquad (3.14)$$

where t is time, β is the distribution shape parameter, and θ is the distribution scale parameter. Obtain an expression for the maintenance worker's reliability.

By substituting Equation (3.14) into Equation (3.13), we get

$$HR(t) = e^{-\int_0^t \left(\frac{\beta t^{\beta-1}}{\theta^\beta} \right) dt}$$

$$= e^{-\left(t/\theta \right)^\beta}$$

(3.15)

Thus, Equation (3.15) is the expression for the maintenance worker's reliability.

3.8.2 HUMAN PERFORMANCE CORRECTABILITY FUNCTION

This is concerned with the human capacity to correct self-generated human errors and is defined as the probability that an error will be corrected in item t subject to stress constraint inherent in the nature of the task and its associated environment [18]. Mathematically, the correctability function is defined as follows [18, 19]:

$$CP(t) = 1 - e^{-\int_0^t \alpha(t) dt}$$

(3.16)

where $CP(t)$ is the probability that an error will be corrected in time t and $\alpha(t)$ is the time-dependent rate at which tasks are corrected.

It is to be noted that Equation (3.16) holds whether the task correction rate is constant or nonconstant. More specifically, it holds for any time to task correction probability distribution.

EXAMPLE 3.3

Assume that the time to error correction of a maintenance worker follows exponential distribution. Thus, his or her error correction rate is defined by

$$\alpha(t) = \alpha$$

(3.17)

where α is the constant error correction rate of the maintenance worker. Obtain an expression for the maintenance worker's correctability function.

Substituting Equation (3.17) into Equation (3.16) yields

$$CP(t) = 1 - e^{-\int_0^t \alpha dt}$$

$$= 1 - e^{-\alpha t}$$

(3.18)

Thus, Equation (3.18) is the expression for the maintenance worker's correctability function.

3.9 HUMAN ERROR OCCURRENCE REASONS, CONSEQUENCES, WAYS, AND CLASSIFICATIONS

Past experiences indicate that there are many reasons for the occurrence of human errors. Some of the important ones are poor training, poor equipment design, poor motivation, complex task, poorly written equipment operating and maintenance procedures, inadequate lighting in the work area, poor management, improper work

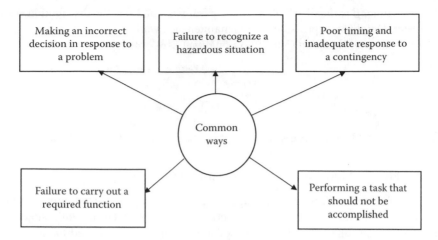

FIGURE 3.2 Common ways human error occurs.

tools, crowded workspace, poor work layout, poor verbal communication, and high noise and temperature in the work area [20].

The consequences of a human error can range from minor to very severe, for example, from insignificant delays in system performance to a very high loss of lives. Furthermore, they may vary from one situation to another, from one task to another, or from one piece of equipment to another. In particular, with respect to equipment, the human error consequences may be grouped under three classifications: equipment operation is stopped completely, equipment operation is delayed quite significantly but not stopped completely, and delay in equipment operation is insignificant.

There are many ways in which a human error can occur. The common ones are shown in Figure 3.2 [21].

Human errors in engineering may be grouped under various classifications. The seven commonly used classifications are as follows [20, 22–24]:

- Maintenance errors
- Operator errors
- Design errors
- Assembly errors
- Inspection errors
- Handling errors
- Contributory errors

Additional information on the above errors is available in Refs. [20, 22–24].

3.10   HUMAN RELIABILITY AND ERROR DATA COLLECTION SOURCES AND QUANTITATIVE DATA

Human reliability and error data are the backbone of any human reliability/error prediction. These data are collected through means such as expert judgments, experimental studies, field experiences, self-made error reports, and published literature [3, 25–26].

TABLE 3.2
Human Reliability and Error Data for Some Selective Tasks

No.	Error/Task Description	Performance Reliability	Error Rate per Million Operations
1	Turning rotary selector switch to certain position	0.9996	—
2	Finding maintenance (scheduled) approaches in maintenance manual	0.997	—
3	Failure to tighten nut and bolt	—	4800
4	Reading gauge incorrectly	—	5000
5	Installing o-ring incorrectly	—	66700
6	Closing valve incorrectly	—	1800
7	Connecting hose incorrectly	—	4700
8	Failure to install nut and bolt	—	600
9	Procedural error in reading instructions	—	64500
10	Incorrect adjustment of mechanical linkage	—	16700

There are many data banks for obtaining human reliability and error-related information [3, 26]. Some of these are Data Store [27], Nuclear Plant Reliability Data System [28], Safety Related Operator Action (SROA) Program [29], Aerojet General Method [30], Bunker-Ramo Tables [31], Air Force Inspection and Safety Center Life Sciences Accident and Incident Reporting System [32], and Aviation Safety Reporting System [33].

Human reliability and error data for some selective tasks, directly or indirectly related to engineering maintenance, are presented in Table 3.2 [3].

3.11 PROBLEMS

1. Discuss three types of man-machine systems.
2. Discuss at least ten comparisons between humans and machines.
3. What are the four main classifications of human factors objectives?
4. List at least six typical human behaviors.
5. Assume that maintenance workers have to read a meter from a distance of 60 inches and the standard character height at a viewing distance of 28 inches is 0.50 inches. Estimate the height of numerals for the stated viewing distance.
6. Discuss at least five sources for collecting human factors-related data.
7. Describe the human performance effectiveness versus stress curve.
8. What are the important reasons for the occurrences of human errors?
9. Discuss five common ways in which a human error can occur.
10. What are the common classifications of human errors in engineering?

REFERENCES

1. MIL-H-46855, Human Engineering Requirements for Military Systems, Equipment, and Facilities, Department of Defense, Washington, D.C., May 1972.
2. Woodson, W.E., *Human Factors Design Handbook*, McGraw-Hill Book Company, New York, 1981.
3. Dhillon, B.S., *Human Reliability: With Human Factors*, Pergamon Press, New York, 1986.
4. Chapanis, A., *Human Factors in Systems Engineering*, John Wiley and Sons, New York, 1996.
5. McCormick, E.J., Sanders, M.S., *Human Factors in Engineering and Design*, McGraw-Hill Book Company, New York, 1982.
6. Dhillon, B.S., *Advanced Design Concepts for Engineers*, Technomic Publishing Company, Lancaster, PA, 1998.
7. AMCP-706-134, Engineering Design Handbook: Maintainability Guide for Design, Prepared by the United States Army Material Command, Alexandria, VA, 1972.
8. Drury, C.G., Fox, J.G., Editors, *Human Reliability in Quality Control*, John Wiley and Sons, New York, 1975.
9. Oborne, D.J., *Ergonomics at Work*, John Wiley and Sons, New York, 1982.
10. Dhillon, B.S., *Engineering Design: A Modern Approach*, Richard D. Irwin, Inc., Chicago, 1996.
11. Peters, G.A., Adams, B.B., Three Criteria for Readable Panel Markings, *Product Engineering*, Vol. 30, 1959, pp. 375–385.
12. Beech, H.R., Burns, L.E., Sheffield, B.F., *A Behavioural Approach to the Management of Stress*, John Wiley and Sons, New York, 1982.
13. Hagen, E.W., Human Reliability Analysis, *Nuclear Safety*, Vol. 17, 1976, pp. 315–326.
14. Meister, D., Human Factors in Reliability, in *Reliability Handbook*, edited by W.G., Ireson, McGraw-Hill Book Company, New York, 1966, pp. 400–415.
15. Dhillon, B.S., On Human Reliability: Bibliography, *Microelectronics and Reliability*, Vol. 20, 1980, pp. 371–373.
16. Regulinski, T.L., Askren, W.B., Mathematical Modeling of Human Performance Reliability, *Proceedings of the Annual Symposium on Reliability*, 1969, pp. 5–11.
17. Askern, W.B., Regulinski, T.L., Quantifying Human Performance for Reliability Analysis of Systems, *Human Factors*, Vol. 11, 1969, pp. 393–396.
18. Regulinski, T.L., Askern, W.B., Stochastic Modeling of Human Performance Effectiveness Functions, *Proceedings of the Annual Reliability and Maintainability Symposium*, 1972, pp. 407–416.
19. Dhillon, B.S., Singh, C., *Engineering Reliability: New Techniques and Applications*, John Wiley and Sons, New York, 1981.
20. Meister, D., The Problem of Human–Initiated Failures, *Proceedings of the 8th National Symposium on Reliability and Quality Control*, 1962, pp. 234–239.
21. Hammer, W., *Product Safety Management and Engineering*, Prentice Hall, Inc., Englewood Cliffs, NJ, 1980.
22. Juran, J.M., Inspector's Errors in Quality Control, *Mechanical Engineering*, Vol. 57, 1935, pp. 643–644.
23. McCormack, R.L., Inspection Accuracy: A Study of the Literature, Report No. SCTM 53–61 (14), Sandia Corporation, Albuquerque, NM, 1961.
24. Meister, D., *Human Factors: Theory and Practice*, John Wiley and Sons, New York, 1971.
25. Meister, D., Human Reliability, in *Human Factors Review*, edited by F.A., Muckler, Human Factors Society, Santa Monica, CA, 1984, pp. 13–53.

26. Dhillon, B.S., Human Error Data Banks, *Microelectrics and Reliability*, Vol. 30, 1990, pp. 963–971.

27. Munger, S.J., Smith, R.W., Payne, D., An Index of Electronic Equipment Operability: Data Store, Report No. AIR-C43-1/62 RP (1), American Institute for Research, Pittsburgh, PA, 1962.

28. Reporting Procedures Manual for the Nuclear Plant Reliability Data System (NPRDS), South-West Research Institute, San Antonio, TX, December 1980.

29. Topmiller, D.A., Eckel, J.S., Kozinsky, E.J., Human Reliability Data Bank for Nuclear Power Plant Operations: A Review of Existing Human Reliability Data Banks, Report No. NUREG/CR2744/1, United States Nuclear Regulatory Commission, Washington, D.C., 1982.

30. Irwin, I.S., Levitz, J.J., Freed, A.M., Human Reliability in the Performance of Maintenance, Report No. LRP317/TDR-63-218, Aerojet General Corporation, Sacramento, CA, 1964.

31. Hornyak, S.J., Effectiveness of Display Sub-Systems Measurement Prediction Techniques, Report No. TR-67-292, Rome Air Development Center (RADC), Griffis Air Force Base, Rome, New York, September 1967.

32. Life Sciences Accident and Incident Classification Elements and Factors, AFISC Operating Instruction No. AFISCM 127-6, United States Air Force, Washington, D.C., December 1971.

33. Aviation Safety Reporting Program, FAA Advisory Circular No. 00-46B, Federal Aviation Administration (FAA), Washington, D.C., June 15, 1979.

4 Methods for Performing Human Reliability and Error Analysis in Engineering Maintenance

4.1 INTRODUCTION

Today, quality, human factors, safety, and reliability are recognized as well-established disciplines. Over the years, many new concepts and methods have been developed in these areas. Many of the methods are being applied quite successfully across many diverse areas including engineering design, production, maintenance, management, and health care. Two important examples of these methods are failure modes and effect analysis (FMEA) and fault tree analysis (FTA).

FMEA was developed by the United States Department of Defense in the early 1950s for analyzing engineering systems from the reliability aspect. Nowadays, FMEA is being used across many diverse areas including maintenance, management, and health care [1–3]. FTA was developed in the early 1960s at the Bell Telephone Laboratories to perform safety and reliability analysis of the Minuteman Launch Control System [3–5]. This method has rapidly gained favor over other reliability and safety analysis methods because of its versatility in degree of detail of complex systems. Today, FTA is being used widely in the industrial sector to analyze problems ranging from management-related to engineering-related.

This chapter presents a number of methods considered useful for performing human reliability and error analysis in engineering maintenance, extracted from the published literature in the areas of quality, human factors, safety, and reliability.

4.2 FAILURE MODES AND EFFECT ANALYSIS (FMEA)

FMEA may simply be described as a powerful method widely used to analyze each potential failure mode in the system under consideration for determining the effects of such modes on the total system [6]. In the event when FMEA is extended to classify each and every potential effect according to its severity, it is called failure mode effects and criticality analysis (FMECA) [7].

The history of FMEA may be traced back to the early 1950s when the United States Navy's Bureau of Aeronautics used it in the design and development of flight control systems [1, 8]. The following main steps are used in performing FMEA [7]:

- **Step 1: Establish system definition.** This is basically concerned with decomposing the system into main blocks and defining their functions, in addition to defining the interface between blocks.
- **Step 2: Establish appropriate ground rules.** This is concerned with formulating the ground rules for performing FMEA. Some examples of these rules are limits of operational stress, statement of primary and secondary mission objectives, delineation of mission phases, limits of environmental stress, and analysis level statement.
- **Step 3: Describe the system and its associated functional blocks.** This is concerned with preparing the description of the system under consideration. This description is normally grouped under two parts:
  - **System block diagram.** This graphically shows the system elements to be analyzed, the system inputs and outputs, series and redundant relationships among the system components/parts, and inputs and outputs of system components.
  - **Functional statement.** This is developed for the total system and for each subsystem and part. The statement is prepared for each operational mode/phase of each item. The degree of detail depends on factors such as the application of the item under consideration and the uniqueness of the function performed.
- **Step 4: Identify possible failure modes and their effects.** This is concerned with systematically identifying the failure modes and their effects. Usually, this is accomplished by using a well-designed worksheet or a form. The worksheet collects data on various areas including item identification and function, failure modes and causes, failure detection approach, failure effects on system/personnel/mission/subsystems, and criticality classification.
- **Step 5: Compile a list of critical items.** This is concerned with developing a list of critical items for providing useful input to sound management decisions. The list contains information on various areas including item identification, concise statement of item's failure mode, classification of criticality, the FMEA worksheet page number, degree of loss effect, and retention rationale.
- **Step 6: Document the analysis.** This is the final step and is concerned with the documentation of analysis. The final document includes items such as system definition and description, ground rules of FMEA, failure modes and their effects, and critical items list.

Some of the important characteristics of the FMEA are as follows:

- By evaluating failure effects of each part, the entire system is screened completely.
- It improves communication quite significantly among individuals involved in the design interface activity.
- It is a routine upward approach that starts from the detail level.

- It highlights weak spots in system design and identifies areas where detailed analysis is necessary.

Additional information on this method is available in Refs. [3, 9].

4.3 MAN–MACHINE SYSTEMS ANALYSIS

This is probably the first method ever developed for reducing human error-caused unwanted effects to some acceptable level, in a system. It was developed in the early 1950s at the Wright-Patterson Air Force Base, United States Air Force, Ohio [10]. The method is composed of the following steps [10].

- **Step 1:** Define the system goals and the associated functions.
- **Step 2:** Define all the situational characteristics; more specifically, the performance shaping factors under which humans will be performing their tasks. Some examples of these factors are quality of air, illumination, and union actions.
- **Step 3:** Define the characteristics (e.g., experience, skills, training, and motivation) of all involved individuals.
- **Step 4:** Define the tasks performed by all involved individuals.
- **Step 5:** Analyze tasks to identify potential error-likely conditions and other associated difficulties.
- **Step 6:** Estimate the chances/other information in regard to the occurrence of each and every potential human error.
- **Step 7:** Estimate the chances that each potential error will remain undetected and uncorrected.
- **Step 8:** Determine the type of consequences if potential human errors remain undetected.
- **Step 9:** Make necessary recommendations for required changes.
- **Step 10:** Reevaluate with care each change by repeating most of the above steps as considered appropriate.

Additional information on this method is available in Ref. [10].

4.4 ROOT CAUSE ANALYSIS (RCA)

RCA may be described as a systematic investigation method that uses data collected during an assessment of an accident, for determining the underlying causes for the deficiencies that led to the occurrence of the accident [11]. As per Ref. [12], RCA was originally developed by the United States Department of Energy.

RCA begins with outlining the event sequence that led to the accident. Starting with the adverse event itself, the analyst involved conducts his or her tasks backward in time, by recording and ascertaining all important events. In collecting such data, it is important for the analyst concerned to avoid making any premature judgment, blame, and attribution, but to specifically focus on the incident-related facts with

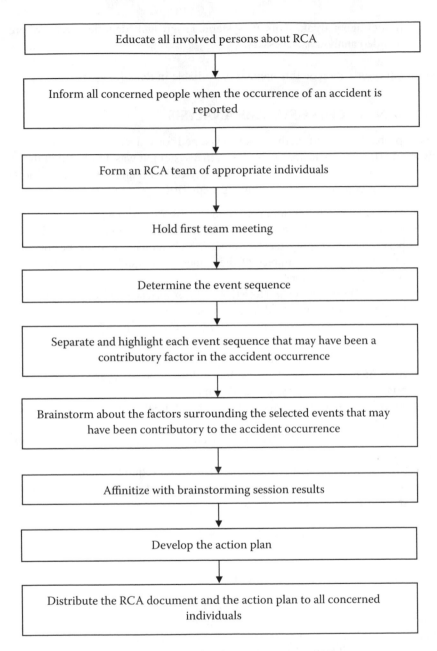

FIGURE 4.1 General steps for performing RCA.

utmost care. Thus, the clearly defined actions leading to an event will be very help-ful to the investigation team members to ask a question with confidence: Why did it (event) occur? [13].

General steps for performing RCA are shown in Figure 4.1 [14]. Additional infor-mation on RCA is available in Refs. [14, 15].

4.5 ERROR-CAUSE REMOVAL PROGRAM (ECRP)

This method was developed specifically for reducing the occurrence of human errors in production operations. The emphasis of the method is on preventive measures rather than merely on remedial ones. Nonetheless, ECRP may simply be described as the production worker participation program to reduce the occurrence of human errors.

The workers who participate in the program include assembly personnel, machin-ists, inspection personnel, maintenance workers, and so on [16]. All these work-ers are grouped under various teams and each team has its own coordinator. The maximum size of the team is twelve workers. Team meetings are held periodically, during which the workers present their error and error-likely reports. The team rec-ommendations are presented to the management for remedial or preventive mea-sures. Usually, teams and management are assisted by various specialists including human factors specialists.

The basic elements of the ECRP are as follows [16, 17]:

- Production workers report and determine errors and error occurrence-likely situations and propose design-related solutions to eradicate error causes.
- Human factors and other specialists evaluate proposed design solutions with respect to cost.
- All people involved with ECRP are educated about its usefulness.
- Management implements the most promising proposed design solutions and recognizes production workers' efforts in an appropriate manner.
- Each worker and team coordinator is properly trained in data collection and analysis approaches.
- The effects of the changes made to the production process are evaluated by human factors and other specialists, with the aid of the ECRP inputs.

Additional information on ECRP is available in Refs. [16, 17].

4.6 CAUSE-AND-EFFECT DIAGRAM (CAED)

This method was developed by a Japanese man named K. Ishikawa in the early 1950s. Occasionally CAED is also called an Ishikawa diagram or a "fishbone dia-gram" because of its resemblance to the skeleton of a fish as shown in Figure 4.2. As shown in the figure, the extreme right-hand side of the diagram (i.e., box or the fish head) represents the effect and the left-hand side represents all the possible causes that are linked to the central line known as the "fish spine."

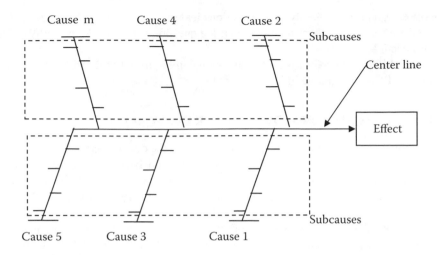

FIGURE 4.2 A cause-and-effect diagram with *m* causes.

In maintenance work, CAED could be a valuable tool to determine the root causes of a given human error-related problem.

The following main steps are used to develop a CAED [18, 19]:

- **Step 1:** Develop problem statement.
- **Step 2:** Brainstorm to identify all possible causes.
- **Step 3:** Develop main cause classifications by stratifying them into natural groups and process steps.
- **Step 4:** Develop the diagram by connecting all the identified causes by following appropriate process steps and fill in the problem or the effect in the diagram box (i.e., the fish head) on the extreme right.
- **Step 5:** Refine the cause classifications by asking questions such as follows:
  - What causes this?
  - What is the real reason for the existence of this condition?

Some of the main benefits of the CAED are that it is a valuable tool to produce ideas, useful to identify root causes, useful to guide further inquiry, and a useful tool for presenting an orderly arrangement of theories [18, 19]. Additional information on CAED is available in Refs [18, 19].

4.7   PROBABILITY TREE METHOD

This method is used to perform task analysis by diagrammatically representing important human actions and other related events. Often, the method is used to perform tasks analysis in the technique for the human error rate prediction (THERP) [20]. In this method, the branches of the probability tree represent diagrammatic task analysis. More specifically, the tree's branching limbs represent the outcome (i.e., success or failure) of each event, and each branch is assigned an appropriate occurrence probability.

Some of the important benefits of the method are as follows [20]:

- A useful visibility tool
- Simplified mathematical computations
- Possesses a good flexibility for incorporating (i.e., with some modifications) factors such as interaction effects, emotional stress, and interaction stress

Additional information on the method is available in Refs. [17, 20]. The following example demonstrates the application of the method.

EXAMPLE 4.1

A maintenance worker performs three independent tasks: x, y, and z. Task x is performed before task y and task y before task z. Each of these three tasks can be performed either correctly or incorrectly. Develop a probability tree and obtain an expression for the probability of not successfully accomplishing the overall mission by the maintenance worker. In addition, calculate the probability of not successfully accomplishing the overall mission by the maintenance worker if the probabilities of performing tasks x, y, and z successfully are 0.8, 0.9, and 0.95, respectively.

In this case, the maintenance worker first performs task x correctly or incorrectly and then proceeds to perform task y. Task y can also be performed correctly or incorrectly. After task y, the worker proceeds to perform task z. This task can also be performed correctly or incorrectly by the maintenance worker. This complete scenario is depicted by Figure 4.3.

The symbols used in Figure 4.3 are defined below.

x denotes the event that task x is performed successfully.
y denotes the event that task y is performed successfully.
z denotes the event that task z is performed successfully.
$\bar{x}$ denotes the event that task x is performed incorrectly.
$\bar{y}$ denotes the event that task y is performed incorrectly.
$\bar{z}$ denotes the event that task z is performed incorrectly.

By examining the diagram, it can be concluded that there are seven distinct possibilities (i.e., $xy\bar{z}, x\bar{y}z, \bar{x}y\bar{z}, \bar{x}yz, x\bar{y}z, x\bar{y}\bar{z}, and \bar{x}\bar{y}\bar{z}$) for not successfully accomplishing the overall mission by the maintenance worker. Thus, the probability of not successfully accomplishing the overall mission by the maintenance worker is expressed by

$$P_{ns} = P(xy\bar{z} + \bar{x}yz + \bar{x}y\bar{z} + \bar{x}\,\bar{y}z + x\bar{y}z + x\bar{y}\,\bar{z} + \bar{x}\,\bar{y}\,\bar{z})$$
$$= P_x P_y P_{\bar{z}} + P_{\bar{x}} P_y P_z + P_{\bar{x}} P_y P_{\bar{z}} + P_{\bar{x}} P_{\bar{y}} P_z + P_x P_{\bar{y}} P_z + P_x P_{\bar{y}} P_{\bar{z}} + P_{\bar{x}} P_{\bar{y}} P_{\bar{z}} \qquad (4.1)$$

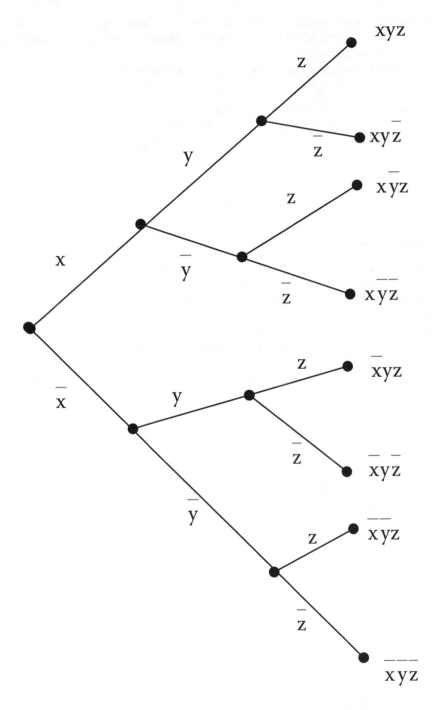

FIGURE 4.3 Probability tree for the maintenance worker performing tasks *x*, *y*, and *z*.

where

P_{ns} is the probability of not successfully accomplishing the overall mission by the maintenance worker.

P_x is the probability of performing task x correctly by the maintenance worker.

P_y is the probability of performing task y correctly by the maintenance worker.

P_z is the probability of performing task z correctly by the maintenance worker.

$P_{\bar{x}}$ is the probability of performing task x incorrectly by the maintenance worker.

$P_{\bar{y}}$ is the probability of performing task y incorrectly by the maintenance worker.

$P_{\bar{z}}$ is the probability of performing task z incorrectly by the maintenance worker.

Because $P_{\bar{x}} = 1 - P_x$, $P_{\bar{y}} = 1 - P_y$, and $P_{\bar{z}} = 1 - P_z$, by substituting the given data values into Equation (4.1), we get

$$P_{ns} = (0.8)(0.9)(1 - 0.95) + (1 - 0.8)(0.9)(0.95) + (1 - 0.8)(0.9)(1 - 0.95)$$

$$+ (1 - 0.8)(1 - 0.9)(0.95) + (0.8)(1 - 0.9)(0.95) + (0.8)(1 - 0.9)(1 - 0.95)$$

$$+ (1 - 0.8)(1 - 0.9)(1 - 0.95)$$

$$= 0.036 + 0.17 + 0.009 + 0.019 + 0.076 + 0.004 + 0.01$$

$$= 0.316$$

Thus, the probability of not successfully accomplishing the overall mission by the maintenance worker is 0.316.

4.8 FAULT TREE ANALYSIS (FTA)

This is a widely used method in the industrial sector for evaluating engineering systems during their design and development phase from reliability and safety aspects. The method was developed in the early 1960s at the Bell Telephone Laboratories by H. A. Watson to perform reliability/safety analysis of the Minuteman Launch Control System [4, 5].

A fault tree may simply be described as a logical representation of the relationship of basic events that lead to a defined undesirable event known as the "top event" and is depicted using a tree structure with logic gates such as AND and OR.

4.8.1 FAULT TREE SYMBOLS

There are many symbols used to construct fault trees of engineering systems. Four of these symbols are shown in Figure 4.4.

The AND gate means that an output fault event occurs only if all the input fault events occur. The OR gate means that an output fault event occurs if one or more input fault events occur. A rectangle represents a fault event that results from the logical combination of fault events through the input of a logic gate such as OR and AND.

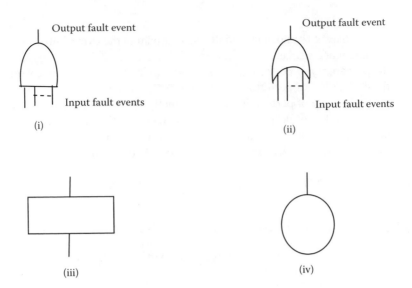

FIGURE 4.4 Four commonly used fault tree symbols: (i) AND gate, (ii) OR gate, (iii) rectangle, (iv) circle.

Finally, a circle denotes a basic fault event or the failure of an elementary part. The fault event's probability of occurrence, failure rate, and repair rate are normally obtained from empirical data. A comprehensive list of fault tree symbols is available in Ref. [21].

4.8.2 STEPS FOR PERFORMING FTA

Usually, the seven steps shown in Figure 4.5 are used to perform FTA [22].

EXAMPLE 4.2

After a careful study of a task being performed by a maintenance worker, it was concluded that he or she can commit an error due to five factors: poor training, inadequate tools, poor instructions, poor environment, or carelessness. Two principal reasons for the poor environment are poor illumination or high noise level. Similarly, two main causes for the poor instructions are poor verbal instructions or poorly written maintenance procedures. Develop a fault tree for the top event "Maintenance worker committed an error" by using the fault tree symbols shown in Figure 4.4.

A fault tree for the example is shown in Figure 4.6. The single capital letters in the diagram denote corresponding fault events (e.g., M: poor environment, N: poor instructions, and A: poor illumination).

4.8.3 PROBABILITY EVALUATION OF FAULT TREES

When the probability of occurrence of basic fault events (e.g., events in circles in Figure 4.6) is given, the probability of occurrence of the top event (e.g., event T in

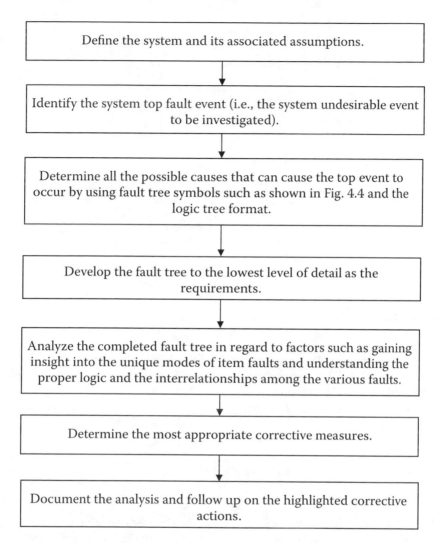

FIGURE 4.5 Steps for performing fault tree analysis (FTA).

Figure 4.6) can be calculated. This can only be calculated by first calculating the probability of occurrence of the output fault events of all the lower and intermediate logic gates (e.g., AND and OR gates).

Thus, the probability of occurrence of the AND gate output fault event, A, is given by [3]

$$P(A) = \prod_{i=1}^{n} P(A_i)$$

(4.2)

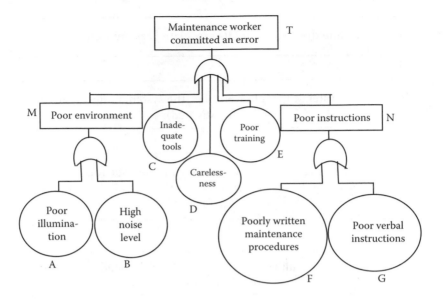

FIGURE 4.6 A fault tree for Example 4.1.

where $P(A)$ is the probability of occurrence of the AND gate output fault event, A; n is the number of AND gate input fault events; and $P(A_i)$ is the occurrence probability of the AND gate input fault event A_i, for $i = 1, 2, 3, n$.

Similarly, the probability of occurrence of the OR gate output fault event, B, is given by [3]

$$P(B) = 1 - \prod_{i=1}^{k} \{1 - P(B_i)\} \qquad (4.3)$$

where $P(B)$ is the probability of occurrence of the OR gate output fault event, B; k is the number of OR gate input fault events; and $P(B_i)$ is the occurrence probability of the OR gate input fault event B_i, for $i = 1, 2, 3, k$.

EXAMPLE 4.3

Assume that the occurrence probabilities of events $A, B, C, D, E, F,$ and G in Figure 4.6 are 0.03, 0.04, 0.05, 0.06, 0.07, 0.08, and 0.09, respectively. Calculate the probability of occurrence of the top event T: maintenance worker committed an error.

By substituting the specified occurrence probability values of the events A and B into Equation (4.3), the probability of the occurrence of event M (i.e., poor environment) is

$$P(M) = 1 - (1 - 0.03)(1 - 0.04)$$
$$= 0.0688$$

Similarly, by substituting the given occurrence probability values of the events F and G into Equation (4.3), the probability of the occurrence of event N (i.e., poor instructions) is

$$P(N) = 1 - (1 - 0.08)(1 - 0.09)$$
$$= 0.1628$$

By substituting the above two calculated values and the given data values into Equation (4.3), we get

$$P(T) = 1 - (1 - 0.0688)(1 - 0.05)(1 - 0.06)(1 - 0.07)(1 - 0.1628)$$
$$= 0.6474$$

where $P(T)$ is the probability of occurrence of the top event T.

Thus, the probability of occurrence of the top event T: maintenance worker committed an error is 0.6474.

4.9   MARKOV METHOD

This is a widely used method in the industrial sector to perform various types of reliability-related studies and is named after the Russian mathematician Andrei Andreyevich Markov (1856–1922). The method is considered quite useful to perform human reliability and error analysis [17]. The following assumptions are associated with the method [22]:

- All occurrences are independent of each other.
- The probability of occurrence of a transition from one state to another in the finite time interval Δt is given by $\alpha \Delta t$, where α is the constant transition rate (e.g., human error rate) from one state to another.
- The transitional probability of two or more occurrences in the finite time interval Δt from one state to another is negligible (e.g., $(\alpha \Delta t)(\alpha \Delta t) \rightarrow 0$).

The following example demonstrates the application of the Markov method, in performing human reliability and error analysis in engineering maintenance.

EXAMPLE 4.4

A maintenance worker is performing a maintenance task on a system used in nuclear power generation. He or she makes errors at a constant rate, α. This scenario is described in more detail by the state space diagram shown in Figure 4.7. The numerals in the circle and box denote system states.

Develop expressions for the maintenance worker's reliability and unreliability at time t and mean time to human error by using the Markov method.

Using the Markov method, we write down the following equations for the diagram [17, 22]:

$$P_0(t + \Delta t) = P_0(t)(1 - \alpha \Delta t) \tag{4.4}$$

$$P_1(t + \Delta t) = P_1(t) + P_0(t)(\alpha \Delta t) \tag{4.5}$$

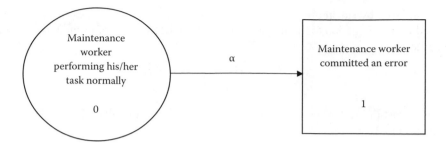

FIGURE 4.7 State space diagram representing the maintenance worker.

where

t is time.

α is the constant error rate of the maintenance worker.

$\alpha \Delta t$ is the probability of human error by the maintenance worker in finite time interval Δt.

$(1 - \alpha \Delta t)$ is the probability of no human error by the maintenance worker in finite time interval Δt.

i is the ith state of the maintenance worker; $i = 0$ means that the maintenance worker is performing his or her task normally, $i = 1$ means that the maintenance worker has committed an error.

$P_i(t)$ is the probability that the maintenance worker is in state i at time t, for $i = 0, 1$.

$P_i(t + \Delta t)$ is the probability that the maintenance worker is in state i at time $(t + \Delta t)$, for $i = 0, 1$.

By rearranging Equations (4.4) and (4.5) and taking the limit as $\Delta t \rightarrow 0$, we obtain

$$\lim_{\Delta t \to 0} \frac{P_0(t + \Delta t) - P_0(t)}{\Delta t} = -\alpha P_0(t) \qquad (4.6)$$

$$\lim_{\Delta t \to 0} \frac{P_1(t + \Delta t) - P_1(t)}{\Delta t} = \alpha P_0(t) \qquad (4.7)$$

Thus, from Equations (4.6) and (4.7), we get

$$\frac{dP_0(t)}{dt} + \alpha P_0(t) = 0 \qquad (4.8)$$

$$\frac{dP_1(t)}{dt} - \alpha P_0(t) = 0 \qquad (4.9)$$

At time $t = 0$, $P_0(0) = 1$ and $P_1(0) = 0$.

By solving Equations (4.8) and (4.9), we obtain

$$P_0(t) = e^{-\alpha t} \tag{4.10}$$

$$P_1(t) = 1 - e^{-\alpha t} \tag{4.11}$$

Thus, expressions for the maintenance worker's reliability and unreliability are given by

$$R_{mw}(t) = P_0(t) = e^{-\alpha t} \tag{4.12}$$

and

$$UR_{mw}(t) = P_1(t) = 1 - e^{-\alpha t} \tag{4.13}$$

where $R_{mw}(t)$ is the maintenance worker's reliability at time t and $UR_{mw}(t)$ is the maintenance worker's unreliability at time t.

The maintenance worker's mean time to human error is given by [17]

$$\begin{aligned}
MTTHE_{mw} &= \int_0^\infty R_{mw}(t)dt \\
&= \int_0^\infty e^{-\alpha t}dt \\
&= \frac{1}{\alpha}
\end{aligned} \tag{4.14}$$

where $MTTHE_{mw}$ is the maintenance worker's mean time to human error.

Thus, expressions for the maintenance worker's reliability and unreliability at time t and mean time to human error are given by Equations (4.12), (4.13), and (4.14), respectively.

EXAMPLE 4.5

A maintenance worker's constant error rate is 0.0009 errors/hour. Calculate his or her unreliability for an 8-hour mission and mean time to human error.

By substituting the specified data values into Equations (4.13) and (4.14), we get

$$\begin{aligned}
UR_{mw}(8) &= 1 - e^{-(0.0009)(8)} \\
&= 0.0072
\end{aligned}$$

and

$$\begin{aligned}
MTTHE_{mw} &= \frac{1}{(0.0009)} \\
&= 1111.1 \text{ hours}
\end{aligned}$$

Thus, the maintenance worker's unreliability and mean time to human error are 0.0072 and 1111.1 hours, respectively.

4.10   PROBLEMS

1. Discuss at least three important characteristics of failure modes and effect analysis.
2. Describe man-machine systems analysis.
3. Compare failure modes and effect analysis with root cause analysis.
4. What are the basic elements of the error-cause removal program?
5. Describe the cause-and-effect diagram. What are its main benefits?
6. A maintenance worker performs two independent tasks: C and D. Task C is performed before task D, and each of these two tasks can be performed either correctly or incorrectly. Develop a probability tree and obtain an expression for the probability of not successfully accomplishing the overall mission by the maintenance worker.
7. What are the main steps for performing fault tree analysis?
8. Describe the following two terms:
   • AND gate
   • OR gate
9. What are the assumptions associated with the Markov method?
10. Prove Equations (4.10) and (4.11) by using Equations (4.8) and (4.9).

REFERENCES

1. Continho, J.S., Failure Effect Analysis, *Transactions of the New York Academy of Sciences*, Vol. 26, Series II, 1963–1964, pp. 564–584.
2. MIL-F-18372 (Aer.), General Specification for Design, Installation, and Test of Aircraft Flight Control Systems, Bureau of Naval Weapons, Department of the Navy, Washington, D.C., Paragraph 3.5.2.3.
3. Dhillon, B.S., *Design Reliability: Fundamentals and Applications*, CRC Press, Boca Raton, FL, 1999.
4. Bennetts, R.G., On the Analysis of Fault Trees, *IEEE Transactions on Reliability*, Vol. 24, No. 3, 1975, pp. 175–185.
5. Dhillon, B.S., Singh, C., Bibliography of Literature on Fault Trees, *Microelectronics and Reliability*, Vol. 17, 1978, pp. 501–503.
6. Omdahl, T.P., Editor, *Reliability, Availability, and Maintainability (RAM) Dictionary*, American Society for Quality Control (ASQC) Press, Milwaukee, WI, 1988.
7. Jordan, W. E., Failure Modes, Effects and Criticality Analysis, *Proceedings of the Annual Reliability and Maintainability Symposium*, 1972, pp. 30–37.
8. Arnzen, H.E., Failure Modes and Effect Analysis: A Powerful Engineering Tool for Component and System Optimization, Report No. 347-40-00-00-K4-05 (C5776), GIDEP Operations Center, United States Navy, Corona, CA, 1966.
9. Palady, P., *Failure Modes and Effect Analysis*, PT Publications, West Palm Beach, FL, 1995.
10. Miller, R.B., A Method for Man-Machine Task Analysis, Report No. 53–137, Wright Air Development Center, Wright-Patterson Air Force Base, U.S. Air Force (USAF), Ohio, 1953.
11. Latino, R.J., Automating Root Cause Analysis, in *Error Reduction in Health Care*, edited by P.L. Spath, John Wiley and Sons, New York, 2000, pp. 155–164.

12. Busse, D.K., Wright, D.J., Classification and Analysis of Incidents in Complex, Medical Environments, Report, 2000. Available from the Intensive Care Unit, Western General Hospital, Edinburgh, UK.

13. Feldman, S.E., Roblin, D.W., Accident Investigation and Anticipatory Failure Analysis in Hospitals, in *Error Reduction in Health Care*, edited by P.L. Spath, John Wiley and Sons, New York, 2000, pp. 139–154.

14. Burke, A., Root Cause Analysis, Report, 2002. Available from the Wild Iris Medical Education, P.O. Box 257, Comptche, CA.

15. Wald, H., Shojania, K.G., Root Cause Analysis, in *Making Health Care Safer: A Critical Analysis of Patient Safety Practices*, edited by A.J. Markowitz, Report No. 43, Agency for Health Care Research and Quality, US Department of Health and Human Services, Rockville, MD, 2001, Chapter 5, pp. 1–7.

16. Swain, A.D., An Error-Cause Removal Program for Industry, *Human Factors*, Vol. 12, 1973, pp. 207–221.

17. Dhillon, B.S., *Human Reliability: With Human Factors*, Pergamon Press, New York, 1986.

18. Ishikawa, K., *Guide to Quality Control*, Asian Productivity Organization, Tokyo, 1982.

19. Mears, P., *Quality Improvement Tools and Techniques*, McGraw-Hill, New York, 1995.

20. Swain, A.D., A Method for Performing a Human-Factors Reliability Analysis, Report No. SCR-685, Sandia Corporation, Albuquerque, NM, August 1963.

21. Dhillon, B.S., Singh, C., *Engineering Reliability: New Technique and Applications*, John Wiley and Sons, New York, 1981.

22. Shooman, M.L., *Probabilistic Reliability: An Engineering Approach*, McGraw-Hill Book Company, New York, 1968.

5 Human Error in Maintenance

5.1 INTRODUCTION

Humans play a pivotal role during system/equipment design, production, operation, and maintenance phases. Although the degree of their role may vary from one phase to another, their interactions are subject to deterioration because of human error. Human error may simply be described as the failure to carry out a given task (or the performance of a forbidden action) that could result in disruption of scheduled operations or damage to equipment and property [1–3].

The occurrence of human error in the maintenance activity can impact equipment performance and safety in various ways. For example, poor repairs can play an instrumental role in increasing the number of equipment breakdowns, which in turn can significantly increase the risk associated with equipment failures and the occurrence of personal accidents [4]. Maintenance error is basically due to wrong preventive actions or repairs, and usually the occurrence of maintenance error increases as the equipment/system ages because of the increase in maintenance frequency.

This chapter presents various important aspects of human error in maintenance.

5.2 FACTS, FIGURES, AND EXAMPLES

Some of the facts, figures, and examples, directly or indirectly, concerned with human error in maintenance are as follows:

- Over 50% of all equipment fail prematurely after the performance of maintenance work [5].
- A study of electronic equipment reported that around 30% of failures were the result of operation and maintenance error [6].
- In 1988, 30 people died and 69 were injured seriously at the Clapham Junction Railway accident in the United Kingdom due to a maintenance error in wiring [7].
- In 1989, the explosion at the Phillips 66 Houston Chemical Complex in Pasadena, Texas, was the result of a maintenance error [8].
- In 1993, a study of 122 maintenance-related occurrences classified maintenance error under four distinct categories: wrong installations (30%), omissions (56%), wrong parts (8%), and miscellaneous (6%) [9, 10].
- A study of an incident that involved the blowout preventer (assembly of valves) at the Ekofisk oil field in the North Sea reported that the incident

was caused by the upside-down installation of the device. The total cost of the incident was estimated to be approximately $50 million [11].

- A study of maintenance tasks such as remove, adjust, and align reported a human reliability mean of 0.9871 [12]. It simply means that management should expect human errors by people involved with the maintenance activity on the order of 13 times in 1000 attempts [11].
- A study of maintenance-related errors in missile operations reported a number of causes: wrong installation (28%), dials and controls (misread, misset) (38%), loose nuts/fittings (14%), inaccessibility (3%), and miscellaneous (17%) [11, 13].

5.3 OCCURRENCE OF MAINTENANCE ERROR IN EQUIPMENT LIFE CYCLE AND ELEMENTS OF A MAINTENANCE PERSON'S TIME

The occurrence of maintenance error during the system/equipment life cycle (i.e., from the time of system/equipment acceptance to the beginning of its phase-out period) is an important factor. Approximate breakdowns of the occurrence of human error in a system/equipment life cycle are shown in Figure 5.1 [11, 14].

A good understanding of time spent by maintenance personnel in performing various maintenance tasks can be quite useful to analyze the occurrence of maintenance errors. Various studies performed over the years indicate that most of their time is spent in the area of fault diagnosis. However, according to one study [11], the maintenance person's time in the area of electronic equipment can be classified

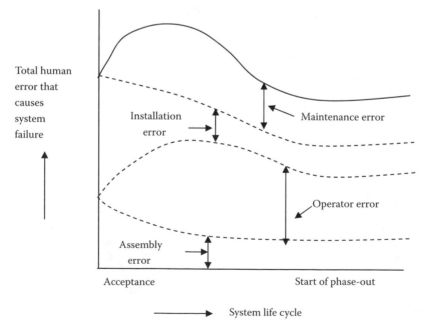

FIGURE 5.1 System life cycle versus four types of human error that cause system failure.

under three categories: diagnosis, remedial actions, and verification. The percentage breakdowns of the time for these three categories are as follows [11]:

- Diagnosis: 65–75%
- Remedial actions: 15–25%
- Verification: 5–15%

5.4 MAINTENANCE ENVIRONMENT AND CAUSES FOR THE OCCURRENCE OF MAINTENANCE ERRORS

As maintenance personnel work directly on equipment, the location of equipment and its design features directly dictate many of the parameters of their work environment. Maintenance environments are susceptible to factors such as noise, poor illumination, and temperature variations. Each of these three factors is described below, separately [15].

5.4.1 NOISE

Maintenance environments can be quite noisy as many are not properly sound-controlled. Ambient noise from ongoing activities can interfere with maintenance personnel's tasks. More specifically, sounds can distract maintenance personnel and interfere with their job performance and sufficiently loud sounds can limit the ability of maintenance personnel to converse or to hear verbal instructions.

Finally, although maintenance personnel can wear protective devices to limit adverse noise effects to a certain degree, these devices can interfere with the performance of their assigned tasks if they are uncomfortable, restrict movement, or hinder conversation.

5.4.2 POOR ILLUMINATION

Lighting deficiencies occur because the external light that maintenance personnel rely on is frequently designed to illuminate the general work area, not the specific areas on which they actually focus. More specifically, illumination-related deficiencies can exist in enclosed or confined spaces, or in places where the primary source of illumination is the overhead lighting.

Finally, maintenance personnel could use portable lighting fixtures to overcome deficiencies such as these; however, if hand-free operations are not possible, their ability to work effectively will be impeded.

5.4.3 TEMPERATURE VARIATIONS

Maintenance personnel may be exposed to wide variations in temperature because they often perform their tasks in outdoor environments or in environments that are not fully climate controlled. Past experiences indicate that maintenance workers and people in general perform effectively at a fairly narrow temperature range.

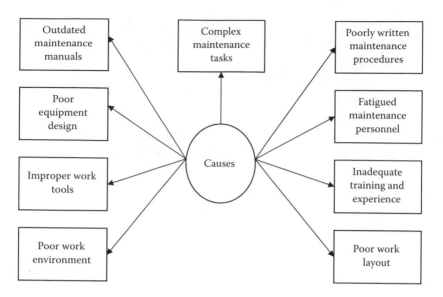

FIGURE 5.2 Causes for the occurrence of maintenance errors.

Furthermore, some studies [15–18] have shown that as the temperature extends beyond a fairly narrow range (i.e., from around 15°C/60°F to about 35°C/90°F), it becomes a stressor that affects the performance of individuals.

Over the years, various studies have identified many different causes for the occurrence of maintenance errors. Some of the important ones are shown in Figure 5.2 [11, 13, 19]. In particular, with regard to training and experience, a study of maintenance personnel reported that those who ranked highest possessed characteristics such as higher aptitude, greater satisfaction with the work group, higher morale, and greater emotional stability [11, 12].

5.5 TYPES OF MAINTENANCE ERRORS AND TYPICAL MAINTENANCE ERRORS

There are basically six types of maintenance errors [5]: recognition failures, memory failures, skill-based slips, knowledge-based errors, rule-based slips, and violation errors.

Recognition failures include items such as nondetection of problem states and misidentification of objects, signals, and messages. Memory failures include items such as input failure (i.e., poor attention is paid to the to-be-remembered item), storage failure (i.e., remembered material decays or suffers interference), premature exit (i.e., terminating a job prior to completing all the necessary actions), and omission following interruptions (i.e., rejoining a sequence of actions and omitting certain necessary steps).

Skill-based slips are usually associated with "automatic" routines and they can include branching errors and overshoot errors. Knowledge-based errors occur when maintenance personnel perform unusual tasks for the first time. Rule-based slips are concerned with misapplying a good rule (i.e., applying a rule in a situation where it is not appropriate) and applying a bad rule (i.e., the rule may get the job/task done under certain conditions, but it can have various consequences).

Finally, violation errors are the deliberate acts which violate procedures. These include thrill-seeking violations (they are frequently committed simply to avoid boredom or win peer praise), routine violations (they are committed to avoid unnecessary effort, get the job/task accomplished quickly, to demonstrate skill acquired, or avoid what is considered as an unnecessarily lengthy procedure/process), and situational violations (they are committed when it is impossible to get the job done if specified procedures are strictly adhered to). Additional information on all the above six types of maintenance errors is available in Ref. [5].

Some of the typical maintenance errors experienced in the industrial sector are as follows [20]:

- Parts installed backward
- Use of incorrect greases, lubricants, or fluids
- Installing incorrect part
- Failure to follow specified procedures and instructions
- Failure to align, check, or calibrate
- Omitting a component or part
- Failure to close or seal properly
- Failure to act on indicators of problems due to factors such as time constraints, priorities, or workload
- Failure to lubricate
- Error resulting from failure to complete task properly because of shift change

5.6 COMMON MAINTAINABILITY DESIGN ERRORS AND USEFUL DESIGN IMPROVEMENT GUIDELINES TO REDUCE EQUIPMENT MAINTENANCE ERRORS

Past experiences indicate that during the equipment design phase often errors are made that adversely affect equipment maintainability and, directly or indirectly, the occurrence of maintenance errors. Some of the common maintainability design errors are as follows [21, 22]:

- Providing poor reliability built-in test equipment
- Placing poor reliability parts beneath other parts
- Placing adjustable screws close to a hot part or an exposed power supply terminal
- Providing inadequate space for maintenance personnel to get their gloved hands into the unit to perform necessary adjustments
- Omitting necessary handles and placing an adjustment out of arm's reach

- Placing adjustable screws in locations difficult for maintenance personnel to find
- Using access doors with numerous small screws and placing screwdriver-related adjustments underneath modules

There are many useful design improvement guidelines for reducing equipment maintenance errors. Some of the important ones are as follows [20]:

- Use operational interlocks in such a way that subsystems cannot be turned on if they are incorrectly assembled or installed.
- Design to facilitate detection of errors and improve warning devices, readouts, and indicators to reduce human decision making.
- Improve fault isolation design by providing appropriate built-in test capability, clearly indicating the direction of fault, and designating test points and procedures.
- Use decision guides to reduce human guesswork by providing appropriate arrows for indicating direction of flow, correct type of fluids/lubricants, and correct hydraulic pressures.
- Improve part-equipment interface by designing interfaces in such a way that the part can only be installed correctly and provide correct mounting pins and other devices for supporting a part/component while it is being bolted or unbolted.

5.7   MAINTENANCE WORK INSTRUCTIONS

Over the years various studies have indicated that omissions account for over 50% of all human factors-related problems in the area of maintenance. Thus, the development and use of effective maintenance work instructions is very essential in managing these types of errors. Some characteristics of good maintenance work instructions are as follows [5]:

- They focus on the risks that may prevent the task/job being carried out safely and to specified quality standards.
- They incorporate sufficient independent inspections at important appropriate points in the instruction.
- They incorporate appropriate and conspicuous reminders for ensuring that important steps are not omitted.
- They group together complex work-related instructions into phases, with each and every phase consisting of many, related tasks/jobs.
- They make use of appropriate pictures and graphics at appropriate places.
- They are written with maintenance personnel who are going to read the instruction in mind.
- They are written clearly and make use of simple and consistent language.

Additional information on the above characteristics is available in Ref. [5].

5.8   MAINTENANCE ERROR ANALYSIS METHODS

Over the years many methods and techniques have been developed to perform various types of analysis in the areas of reliability, quality, and safety. Some of these methods can also be used to perform maintenance error analysis. Four of these methods are presented below.

5.8.1   Probability Tree Method

This is one of the commonly used methods to perform human reliability analysis. It is considered a quite useful approach to perform task analysis in maintenance work. In performing task analysis, the approach diagrammatically represents human actions. Thus, diagrammatic task analysis is denoted by the probability tree branches.

More specifically, the branching limbs denote outcomes (i.e., success or failure) of each event or action associated with a problem under consideration. Also, each branch of the probability tree is assigned an occurrence probability.

The method is described in detail in Chapter 4 and in Refs. [13, 21]. Its application to performing maintenance error analysis is demonstrated through the example presented below.

Example 5.1

Assume that a maintenance person performs two independent tasks, say, m and n. Task m is performed before task n and each of these two tasks can be either performed correctly or incorrectly. Draw the probability tree for the example and obtain probability expressions for the following:

1. Successfully accomplishing the overall mission by the maintenance person.
2. Not successfully accomplishing the overall mission by the maintenance person.

In this case, the maintenance person first performs task m correctly or incorrectly and then proceeds to performing task n. This complete scenario is represented by the probability tree diagram in Figure 5.3.

The four symbols used in Figure 5.3 are defined below.

m denotes the event that task m is performed correctly by the maintenance person.

$\bar{m}$ denotes the event that task m is performed incorrectly by the maintenance person.

n denotes the event that task n is performed correctly by the maintenance person.

$\bar{n}$ denotes the event that task n is performed incorrectly by the maintenance person.

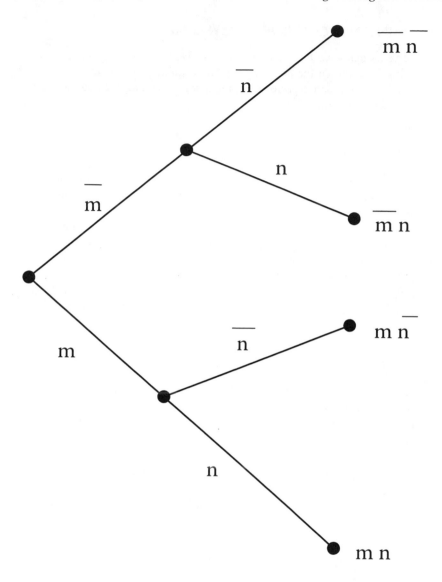

FIGURE 5.3 Probability tree for the maintenance person performing tasks *m* and *n*.

By examining the diagram, it can be noted that there are three distinct possibilities (i.e., $\overline{m}\,\overline{n}$, $\overline{m}n$, *and* $m\overline{n}$) for not successfully accomplishing the overall mission by the maintenance person. Thus, the probability of not successfully accomplishing the overall mission by the maintenance person is given by

$$P_f = P(\overline{m}\,\overline{n} + \overline{m}n + m\overline{n})$$
$$= P_{\overline{m}}P_{\overline{n}} + P_{\overline{m}}P_n + P_m\,P_{\overline{n}} \tag{5.1}$$

where

P_f is the probability of not successfully accomplishing the overall mission by the maintenance person.

P_m is the probability of performing task m correctly by the maintenance person.

P_n is the probability of performing task n correctly by the maintenance person.

$P_{\bar{m}}$ is the probability of performing task m incorrectly by the maintenance person.

$P_{\bar{n}}$ is the probability of performing task n incorrectly by the maintenance person.

Because $P_{\bar{m}} = 1 - P_m$ and $P_{\bar{n}} = 1 - P_n$, Equation (5.1) reduces to

$$P_f = \left(1 - P_m\right)\left(1 - P_n\right) + \left(1 - P_m\right)P_n + P_m\left(1 - P_n\right)$$
$$= 1 - P_n P_m$$

(5.2)

Similarly, by examining Figure 5.3, it can be noted that there is only one possibility (i.e., mn) for successfully accomplishing the overall mission by the maintenance person. Thus, the probability of successfully accomplishing the overall mission by the maintenance person is given by

$$P_s = P(mn)$$
$$= P_m P_n$$

(5.3)

where P_s is the probability of successfully accomplishing the overall mission by the maintenance person.

EXAMPLE 5.2

Assume that in Example 5.2, the probabilities of the maintenance person performing tasks m and n correctly are 0.9 and 0.95, respectively. Calculate the probability of not successfully accomplishing the overall mission by the maintenance person.

By substituting the given data values into Equation (5.2), we get

$$P_f = 1 - (0.95)(0.9)$$
$$= 0.855$$

Thus, the probability of not successfully accomplishing the overall mission by the maintenance person is 0.855.

5.8.2 PONTECORVO METHOD

This is a quite useful method that can be used to obtain reliability estimates of task performance by a maintenance person. The method first obtains reliability estimates for separate and discrete subtasks having no correct reliability figures, and then it combines these estimates to obtain the total task reliability. Usually, the Pontecorvo approach is applied during initial design phases and is composed of the six steps shown in Figure 5.4 [13, 22].

Step 1 is concerned with the identification of tasks to be performed. These tasks are to be identified at a gross level (i.e., each task is represented by one complete operation). Step 2 is concerned with the identification of those subtasks that are essential for task completion. Step 3 is concerned with collecting data from sources such as in-house operations and experimental literature.

Step 4 is concerned with rating each subtask according to its potential for error or level of difficulty. Normally, a 10-point scale is used to judge the appropriate subtask rate. The scale varies from least error to most error. Step 5 is concerned with predicting the subtask reliability and is accomplished by expressing the judged ratings of the data and the empirical data in the form of a straight line. The regression line is tested for goodness of fit.

Finally, Step 6 is concerned with determining the task reliability. The task reliability is obtained by multiplying reliabilities of all the subtasks.

It is to be noted that the above approach is used to estimate the performance of a single individual acting alone. However, when a backup person is available, the probability of the task being performed correctly (i.e., the task reliability) improves. Nonetheless, the backup individual may not be available all of the time. In such a scenario, the overall reliability of two individuals working together to accomplish a specified task can be estimated by utilizing the following expression [13, 22]:

$$R_O = \left[\left\{ 1 - (1 - R_s)^2 \right\} PT_1 + R_s PT_2 \right] / (PT_1 + PT_2) \qquad (5.4)$$

where R_s denotes the single person reliability, PT_1 denotes the percentage of time the backup person is available, and PT_2 denotes the percentage of time the backup person is unavailable.

EXAMPLE 5.3

Two maintenance workers are working independently together to carry out a maintenance-related task. The reliability of each worker is 0.90, and the backup worker is only available 40% of the time. Calculate the reliability of performing the maintenance task correctly.

Thus, as per the specified data value, the percentage of time the backup maintenance worker is unavailable is given by

$$PT_2 = 1 - PT_1$$
$$= 1 - 0.40$$
$$= 0.60 \ or \ 60\%$$

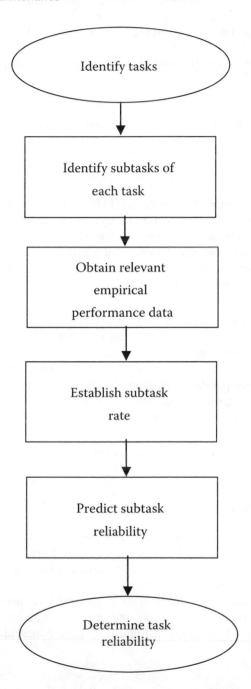

FIGURE 5.4 Pontecorvo method steps.

Using the above calculated value and the given data values in Equation (5.4), we get

$$R_O = [\{1 - (1 - 0.9)^2\}0.4 + (0.9)(0.6)] / (0.4 + 0.6)$$
$$= 0.936$$

Thus, the reliability of carrying out the maintenance task correctly is 0.936.

5.8.3 PARETO ANALYSIS

The method is named after Vilfredo Pareto (1848–1923), an Italian economist, and it is a quite useful method that can be used to separate the important causes of maintenance error-related problems from the trivial ones.

Thus, the method is considered a powerful tool to identify areas for a concerted effort to minimize or eliminate the occurrence of maintenance errors. The method is composed of the six steps listed below [23, 24].

- **Step 1:** List causes in tabular form and count their occurrences.
- **Step 2:** Arrange the causes in descending order.
- **Step 3:** Calculate the total for the entire list.
- **Step 4:** Determine the percentage of the total for each cause.
- **Step 5:** Develop a Pareto diagram that shows percentages vertically and their corresponding causes horizontally.
- **Step 6:** Conclude from the final results.

Additional information on Pareto analysis is available in Refs. [23, 24].

5.8.4 MARKOV METHOD

This is a widely used tool to perform various types of reliability analysis, and it can be used to perform human error analysis in maintenance work. The method is described in Chapter 4. Its application in the area of maintenance is demonstrated through the following mathematical model.

This mathematical model represents a maintenance person performing a maintenance task. He or she can make and self-correct an error. The state space diagram of the model is shown in Figure 5.5 [24]. Numerals in boxes denote system states.

The model is subject to the following assumptions:

- The maintenance person's error and self-error-correction rates are constant.
- The maintenance person can self-correct his or her errors.
- After the error correction the maintenance person's performance remains normal.

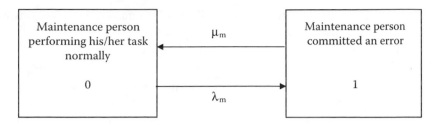

FIGURE 5.5 State space diagram for the maintenance person.

The following symbols are associated with the model:

i is the maintenance person's state; for $i = 1$ (maintenance person performing his or her task normally), $i = 1$ (maintenance person committed an error).

$P_i(t)$ is the probability that the maintenance person is in state i at time t; for $i = 0,1$.

λ_m is constant error rate of the maintenance person.

μ_m is constant self-error-correction rate of the maintenance person.

With the aid of the Markov method, we write down the following equations for the diagram:

$$\frac{dP_0(t)}{dt} + \lambda_m P_0(t) = P_1(t)\mu_m \tag{5.5}$$

$$\frac{dP_1(t)}{dt} + \mu_m P_1(t) = P_0(t)\lambda_m \tag{5.6}$$

At time $t = 0$, $P_0(0) = 1$ and $P_1(0) = 0$.

Solving Equations (5.5) and (5.6), we get

$$P_0(t) = \frac{\mu_m}{(\lambda_m + \mu_m)} + \frac{\lambda_m}{(\lambda_m + \mu_m)} e^{-(\lambda_m + \mu_m)t} \tag{5.7}$$

$$P_1(t) = \frac{\lambda_m}{(\lambda_m + \mu_m)} - \frac{\lambda_m}{(\lambda_m + \mu_m)} e^{-(\lambda_m + \mu_m)t} \tag{5.8}$$

As time t becomes very large, we get the following steady-state probability equations from Equations (5.7) and (5.8), respectively:

$$P_0 = \frac{\mu_m}{\lambda_m + \mu_m} \tag{5.9}$$

$$P_1 = \frac{\lambda_m}{\lambda_m + \mu_m} \tag{5.10}$$

where P_0 and P_1 are the steady-state probabilities of the maintenance person being in states 0 and 1, respectively.

EXAMPLE 5.4

A maintenance person is performing a maintenance task and his or her error and self-error-correction rates are 0.0003 errors/hour and 0.0001 errors/hour, respectively. Calculate the maintenance person's probability of correctly performing his or her task during an 8-hour period.

By substituting the specified data values into Equation (5.7), we get

$$P_0(8) = \frac{0.0001}{(0.0003 + 0.0001)} + \frac{(0.0003)}{(0.0003 + 0.0001)} e^{-(0.0003 + 0.0001)(8)}$$
$$= 0.9976$$

Thus, the maintenance person's probability of performing his or her task correctly is 0.9976.

5.9   PROBLEMS

1. Give at least four facts and figures concerned with human error in maintenance.
2. Discuss the occurrence of maintenance error in equipment life cycle.
3. Write an essay on the maintenance environment.
4. What are the main causes for the occurrence of maintenance errors?
5. What are the six basic types of maintenance errors?
6. List at least eight typical maintenance errors.
7. What are the common maintainability design errors?
8. Discuss maintenance work instructions.
9. Describe the following two items:
   • Pareto analysis
   • Pontecorvo method
10. Prove Equations (5.7) and (5.8) by using Equations (5.5) and (5.6).

REFERENCES

1.   Meister, D., The Problem of Human-Initiated Failures, *Proceedings of the 8th National Symposium on Reliability and Quality Control*, 1962, pp. 234–239.
2.   Meister, D., Human Factors in Reliability, in *Reliability Handbook,* edited by W. G. Ireson, McGraw-Hill, New York, 1966, pp. 12.2–12.37.
3.   Hagen, E.W., Human Reliability Analysis, *Nuclear Safety,* Vol. 17, 1976, pp. 315–326.
4.   Mason, S., Improving Maintenance by Reducing Human Error, 2007. Available from Health Safety and Engineering Consultants Ltd., 70 Tamworth Road, Ashby-de-la-Zouch, Leicestershire, UK.
5.   Dunn, S., Managing Human Error in Maintenance, 2007. Available from Assetivity Pty Ltd., P.O. Box 1315, Boorgoon, WA 6154.
6.   AMCP 706–134, Maintainability Guide for Design, U.S. Army Material Command, Department of the Army, Washington, D.C., 1972.

7. Report: Investigation into the Clapham Junction Railway Accident, Department of Transport, Her Majesty's Stationery Office, London, UK, 1989.
8. Reason, J., Hobbs, A., *Managing Maintenance Error: A Practical Guide,* Ashgate Publishing Company, Aldershot, UK, 2003.
9. Circular 243-AN/151, Human Factors in Aircraft Maintenance and Inspection, International Civil Aviation Organization, Montreal, Canada, 1995.
10. Human Factors in Airline Maintenance: A Study of Incident Reports, Bureau of Air Safety Investigation, Department of Transport and Regional Development, Canberra, Australia, 1997.
11. Christensen, J.M., Howard, J.M., Field Experience in Maintenance, in *Human Detection and Diagnosis of System Failures,* edited by J. Rasmussen and W.B. Rouse, Plenum Press, New York, 1981, pp. 111–133.
12. Sauer, D., Campbell, W.B., Potter, N.R., Askern, W.B., Relationships between Human Resource Factors and Performance on Nuclear Missile Handling Tasks, Report No. AFHRL-TR-76-85/AFWL-TR-76-301, Air Force Human Resources Laboratory/Air Force Weapons Laboratory, Wright-Paterson Air Force Base, Ohio, 1976.
13. Dhillon, B.S., *Human Reliability: With Human Factors*, Pergamon Press, New York, 1986.
14. Rigby, L.V., The Sandia Human Error Rate Bank (SHERB), Report No. SC-R-67-1150, Sandia Laboratories, Albuquerque, NM, 1967.
15. Strauch, B., *Investigating Human Error: Incidents, Accidents, and Complex Systems*, Ashgate Publishing Limited, Aldershot, UK, 2002.
16. Ellis, H.D., The Effects of Cold on the Performance of Serial Choice Reaction Time and Various Discrete Tasks, *Human Factors,* Vol. 24, 1982, pp. 589–598.
17. Van Orden, K.F., Benoit, S.L., Osga, G.A., Effects of Cold Air Stress on the Performance of a Command and Control Task, *Human Factors*, Vol. 38, 1996, pp. 130–141.
18. Wyon, D.P., Wyon, I., Norin, F., Effects of Moderate Heat Stress on Driver Vigilance in a Moving Vehicle, *Ergonomics*, Vol. 39, 1996, pp. 61–75.
19. Dhillon, B.S., *Engineering Maintenance: A Modern Approach*, CRC Press, Boca Raton, FL, 2002.
20. Under, R.L., Conway, K., Impact of Maintainability Design on Injury Rates and Maintenance Costs for Underground Mining Equipment, in *Improving Safety at Small Underground Mines*, Compiled by R.H. Peters, Special Publication No. 18–94, Bureau of Mines, United States Department of the Interior, Washington, D.C., 1994.
21. Dhillon, B.S., Singh, C., *Engineering Reliability: New Techniques and Applications*, John Wiley and Sons, New York, 1981.
22. Pontecorvo, A.B., A Method of Predicting Human Reliability, *Proceedings of the 4th Annual Reliability and Maintainability Conference*, 1965, pp. 337–342.
23. Kanji, G.K., Asher, M., *100 Methods for Total Quality Management,* Sage Publications London, 1996.
24. Dhillon, B.S., *Design Reliability: Fundamentals and Applications*, CRC Press, Boca Raton, FL, 1999.

6 Human Factors in Aviation Maintenance

6.1 INTRODUCTION

An efficient and safe air travel system depends basically on three elements: design, operation, and maintenance. Each year a vast sum of money is spent on aviation maintenance throughout the world. For example, according to the United States Air Transport Association, U.S. airlines spend around $9 billion on maintenance each year [1]. This represents roughly 12% of the total operating cost of an airline company.

Aviation maintenance has changed over the years because newer aircraft contain power plants, electronic subsystems, and materials that did not exist in earlier models [2, 3]. In turn, aircraft maintenance personnel are using increasingly sophisticated equipment and procedures. However, one important aspect of aviation maintenance that has not changed is that most maintenance tasks are still being performed by human inspectors and technicians.

Needless to say, although the aircraft on which these maintenance personnel work have evolved dramatically over the past 50 years, the maintenance personnel still exhibit all of the limitations, idiosyncrasies, and capabilities that are part of being human.

This chapter presents various important aspects of human factors in aviation maintenance.

6.2 THE NEED FOR HUMAN FACTORS IN AVIATION MAINTENANCE AND HOW HUMAN FACTORS IMPACT AIRCRAFT ENGINEERING AND MAINTENANCE

According to the Annual Report of the United States, scheduled Airline Industry, costs, passenger miles flown, and number of aircraft have all exceeded the overall growth of the aviation maintenance technician (AMT) workforce over a period of ten years (i.e., 1983–1993) [4]. It simply means that AMT must enhance efficiency to match the increasing workload demanded by the combination of new skill and knowledge requirements for advanced technology aircraft and increasing labor demand appropriate for providing continuing airworthiness to the existing fleet.

In order to achieve these goals effectively, individual technician's skills and responsibilities must increase to a significant level. Moreover, the airline industry and agencies such as the United States Federal Aviation Administration (FAA) must strive to ensure that maintenance personnel become better qualified and that maintenance works and procedures become more simplified.

Past experiences indicate that human factors can impact aircraft engineering and maintenance in many different ways [5]. For example, at the design and manufacturing stage, critical parts must be identified and manufactured according to the requisite standards. Subsequently, these parts must be subject to inspection and test requirements, as appropriate, in the aircraft maintenance schedule. If they are not in the schedule, then the planning engineer cannot be blamed for not calling a check. Similarly, the aircraft engineer cannot be blamed for overlooking to perform an inspection that was not called for, unless the fault is very obvious.

Nonetheless, engineering designers can take various steps to minimize the occurrence of certain maintenance errors. Two examples of these steps are making critical part areas readily inspectable and devising appropriate checkout procedures to cater for maintenance errors which could cause hazards.

Other human factors that can have a direct effect on aircraft engineering and maintenance include pressure and stress (i.e., either actual or perceived), environment (e.g., too dark, too cold), and circadian rhythm (i.e., natural body variations on shift work) [5].

6.3 HUMAN FACTORS CHALLENGES IN AVIATION MAINTENANCE

There are many human factors challenges in aviation maintenance. The primary challenges can be identified under five classifications as shown in Figure 6.1 [6].

The classification *the worker* is basically concerned with the availability of adequately qualified aviation maintenance personnel in the future. The classification *the workplace* is concerned with providing an effective workplace to aviation maintenance personnel with respect to factors such as safety, temperature, lighting, work access, and noise. The classification *training* is basically concerned with continuously providing proper training to aviation maintenance personnel with respect to changing aircraft-related technologies.

The classification *communication* is concerned with providing timely and accurate maintenance task performance information to aviation maintenance personnel with respect to factors such as "user-friendly" manuals, work cards, and other sources for obtaining inspection and repair-related information. Finally, the classification *aircraft systems* is concerned with, in addition to considering the traditional maintainability-related factors during aircraft systems design, the specific needs of aviation maintenance and inspection manpower at the initial stage of the aircraft systems design.

Additional information on these five classifications of human factors challenges is available in Ref. [6].

6.4 PRACTICAL HUMAN FACTORS GUIDE FOR THE AVIATION MAINTENANCE ENVIRONMENT

Aviation maintenance personnel work as one element within the framework of a large industrial system that contains elements such as the maintenance facility, aircraft, inspection equipment, repair equipment, and supervisory forces [7]. In order

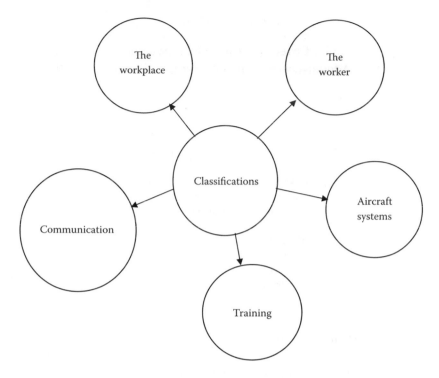

FIGURE 6.1 Classifications of human factors challenges in aviation maintenance.

to understand the performance of maintenance personnel working within the frame-work of this system, proper information is needed concerning the operating characteristics of this very element, that is, maintenance personnel. Two examples of the required information are the following:

- How do the maintenance personnel work?
- What features of the maintenance personnel and/or the environment tend to generate maintenance error?

Human factors is a discipline that seeks to provide appropriate answers to questions such as listed above through an understanding of factors such as the capabilities and limitations of humans, human behavior laws, and the possible effects of the environment on human performance. Thus, a goal of human factors is to draw on knowledge of these factors in developing guidelines for the optimum use of humans in operating systems.

In order to meet this objective the FAA has developed a guidebook titled *Human Factors in Aviation Maintenance* that presents human factors information oriented specifically towards their carrier maintenance personnel. The guidebook contains 12 chapters on 12 different topics, listed in Table 6.1 [3].

TABLE 6.1
Topics Covered in the Federal Aviation
Administration Human Factors Guidebook

No.	Topic
1	Human factors
2	Facility design
3	Establishing human factors/ergonomics program
4	Workplace and job design
5	Workplace safety
6	Training
7	Testing and troubleshooting
8	Automation
9	Shift work and scheduling
10	Personal and job-related factors
11	Sexual harassment
12	Disabilities

The chapter "Human Factors" introduces the field of human factors/ergonomics and defines various concepts and terminology related to factoring human capabilities and limitations into the workplace environment. The chapter "Facility Design" describes the important human factors concepts concerning designing facilities by emphasizing the elements found in the aviation maintenance environment. Two examples of these elements are movable scaffolding and large open hanger areas.

The chapter "Establishing Human Factors/Ergonomics Program" discusses what an ergonomics program is and why an aviation organization should have one. The chapter also describes items such as the concept of a systematic framework for ensuring that human factors are properly considered in the maintenance organization, the regulatory requirements that relate to human factors programs, and the steps required to set up a human factors program. The chapter "Workplace and Job Design" describes the human factors concepts underlying the proper design of jobs and workplaces and the recent research efforts in the aviation maintenance environment that is part of the ongoing FAA emphasis.

The chapter "Workplace Safety" discusses items such as the major hazards associated with industrial workplaces, the steps maintenance supervisors and planners should take to mitigate the hazards, and the features unique to the aviation maintenance workplace.

The chapter "Training" describes various important items concerning training including the overall training requirements in the aviation maintenance environment, changes to training required from the regulatory perspective, and the training methods appropriate for teaching various types of knowledge and skills.

The chapter "Testing and Troubleshooting" discusses the human factors concepts and methods that relate, directly or indirectly, to aviation maintenance testing and troubleshooting. The chapter "Automation" describes the most useful concepts concerning automation, that is, both in general and in the aviation maintenance environment. More specifically, it describes how to decide which maintenance functions are most amenable to automation, as well as various myths and potential automation pitfalls.

The chapter "Shift Work and Scheduling" discusses important research findings concerning various shift scheduling practices including the concepts of circadian rhythms, desynchronization, and the effects of sleep deprivation. The chapter "Personal and Job-Related Factors" discusses issues such as job-related stress, financial concerns, substance abuse, and family problems along with the proper use and potential misuse of employee assistance programs.

The chapter "Sexual Harassment" discusses various aspects of sexual harassment including the underlying social and legal concepts concerning sexual harassment, and the latest court decisions and regulatory requirements. Finally, the chapter "Disabilities" describes the requirements of the Americans with Disabilities Act (ADA) and its implications for the aviation maintenance environment, along with a human factors perspective on adjusting to the capabilities and limitations of people with disabilities.

Additional information on all of the above twelve topics is available in Ref. [3].

6.5 INTEGRATED MAINTENANCE HUMAN FACTORS MANAGEMENT SYSTEM (IMMS)

IMMS is the ongoing European effort toward the integrated management of human factors in aircraft maintenance. More specifically, it is the part of the HILAS (Human Integration into the Lifecycle of Aviation Systems) project divided into four parallel strands of work: the monitoring and assessment of maintenance operations, the integration and management of human factors knowledge, the evaluation of new flight deck technologies, and the flight operations environment and performance [8].

Some of the main objectives of the IMMS are to improve operational performance, improve safety performance, reduce human factors-related risks, and improve quality.

There are five main components (i.e., C_1, C_2, C_3, C_4, and C_5) of the IMMS divided into two categories: front applications and back applications. Thus, the front and back application components are C_1, C_2 and C_3, C_4 and C_5, respectively. Each of these five components is described below [8].

- C_1: This is for aircraft maintenance engineers and it will provide these engineers better task support, through a portable handheld device, employing modern technologies such as radio frequency identification (RFID) and virtual reality.
- C_2: This is for all the support functions. More specifically, it will both provide information to these support functions on how to manage the "softer" aspects of managing the checks and any difficulty experienced.
- C_3: This will collect data from the front applications (i.e., C_1 and C_2) in addition to collecting data from currently operating systems such as

planning, engineering, and quality systems, within the organization. Also, C_3 will allow all these systems to communicate with each other.

- **C_4:** This is the suite of Human Factor Tools and Methods that will manage the human component of the system. Directly or indirectly, the data from components C_1, C_2, and C_3 will continuously update this component (i.e., C_4).
- **C_5:** This deals with implementation on two levels. The first level is concerned with implementing the actual recommendations that come out of the system, whereas the second level is concerned with the implementation of the system itself. This component, i.e., C_5, will also address the wider issue of organizational support.

Additional information on IMMS is available in Ref. [8].

6.6 AVIATION MAINTENANCE HUMAN FACTORS TRAINING PROGRAM AND HUMAN FACTORS TRAINING AREAS FOR AVIATION MAINTENANCE PERSONNEL

One of the most challenging issues in aviation maintenance is designing and developing appropriate human factors training programs. A systematic method that can be used to design and develop human factors training programs is composed of five processes/steps as shown in Figure 6.2 [9–12]. This process includes items such as

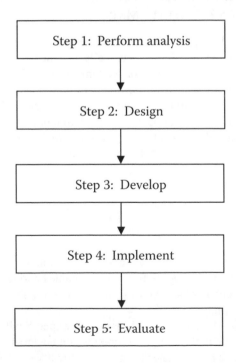

FIGURE 6.2 Steps of the method for designing and developing human factors training programs.

establishing goals and defining training objectives, developing and implementing the training program, involving end users and/or subject matter experts, measuring the training effects, and providing feedback to the training developers [9]. The steps shown in Figure 6.2 are described below [9].

- **Step 1: Perform analysis:** This is concerned with performing three types of analysis: organizational, task, and person. The purpose of these analyses is to determine the degree of training needs and performance gaps, develop hierarchical task analyses, and create an appropriate learning hierarchy which identifies the existing knowledge, skills, and ability levels of trainees under consideration.
- **Step 2: Design:** This is concerned with defining the instructional curriculum, goals, and objectives. This can be accomplished quite effectively by adopting a participatory design approach that includes the creation of a multidisciplinary team of experts, end users, in the areas of aviation maintenance, maintenance operations, inspections, FAA regulations, and human factors.
- **Step 3: Develop:** This is concerned with developing the training materials and media. The incorporation of "in-house" examples in such materials could be very useful.
- **Step 4: Implement:** This is concerned with implementing and delivering the training to trainees.
- **Step 5: Evaluate:** This is concerned with evaluating the training and it includes measuring the effectiveness of the training program on the trainees' performance, behaviors, and knowledge. There are a number of approaches used to evaluate training courses based on a five-level framework [11, 12, 13–15]. These five levels are baseline assessment prior to training, trainee reaction, learning, performance (i.e., behavioral changes), and organizational results [9].

There are many disciplines of human factors including educational psychology, organizational psychology, cognitive science, safety engineering, clinical psychology, experimental psychology, and anthropometric engineering [16]. Because of these many disciplines, aviation maintenance human factors courses can have many approaches with varying instructional goals. Nonetheless, some of the useful topics/areas as candidates for an aviation maintenance human factors course are as follows [16]:

- Safety and economic statistics
- Error and error reporting with respect to economics of error and corporate/regulatory discipline
- Maintenance crew resource management
- Stress
- Human factors fundamentals including analytic methods, human performance models, environmental factors, physical factors, medical factors and health, and cognitive factors
- Teamwork

- Workplace safety
- Behavioral analysis
- Psychological factors
- Situation awareness
- Communication in the workplace that includes items such as principles of communication, leadership, conflict resolution, decision making, planning meetings, and group dynamics/teamwork.

6.7 COMMON HUMAN FACTORS–RELATED AVIATION MAINTENANCE PROBLEMS

Past experiences indicate that there are many human factors–related aviation maintenance problems. Some of the common ones are as follows [17]:

- Coexistence of various types of new technology and old technology equipment
- Need for appropriate advanced technology job aids
- Availability of qualified manpower
- Need for effective technical training for troubleshooting
- Need for task analytic job performance-related data
- Suboptimal working conditions/environments
- Organization and effective usability of all technical documentation

Additional information on these common human factors–related aviation maintenance problems is available in Ref. [17].

6.8 PROBLEMS

1. Discuss the need for human factors in aviation maintenance.
2. List at least six common aviation maintenance problems with respect to human factors.
3. Describe the Integrated Maintenance Human Factors Management System (IMMS).
4. List at least ten topics covered in the Federal Aviation Administration (FAA) human factors guidebook for the aviation maintenance environment.
5. Discuss how human factors impact aircraft engineering and maintenance.
6. What are the major human factors challenges in aviation maintenance?
7. Describe the steps of the method for designing and developing human factors training programs.
8. List at least 10 topics/areas as prime candidates for an aviation maintenance human factors training course.
9. Write a short essay on human factors in aviation maintenance.
10. Discuss the need for having a practical human factors guide for the aviation maintenance environment.

REFERENCES

1. Johnson, W.B., Human Factors Research: Can it Have an Impact on a Financially Troubled U.S. Aviation Industry? *Proceedings of the Human Factors and Ergonomics Society 37th Annual Meeting*, 1993, pp. 21–25.
2. Vreeman, J., Changing Air Carrier Maintenance Requirements, *Proceedings of the Sixth Meeting on Human Factors Issues in Aircraft Maintenance and Inspection*, 1992, pp. 40–48.
3. Maddox, M.E., Introducing a Practical Human Factors Guide into the Aviation Maintenance Environment, *Proceedings of the Human Factors and Ergonomics Society 38th Annual Meeting*, 1994, pp. 101–105.
4. Shepherd, W.T., Human Factors in Aviation Maintenance and Inspection: Research Responding to Safety Demands of Industry, *Proceedings of the Human Factors and Ergonomics Society 39th Annual Meeting*, 1995, pp. 61–65.
5. Nunn, R., Witts, S.A., The Influence of Human Factors on the Safety of Aircraft Maintenance, *Proceedings of the 50th Flight Safety Foundation/International Federation of Airworthiness Conference*, 1997, pp. 211–221.
6. Shepherd, W.T., Human Factors Challenges in Aviation Maintenance, *Proceedings of the Human Factors Society 36th Annual Meeting*, 1992, pp. 82–86.
7. Parker, J.F., A Human Factors Guide for Aviation Maintenance, *Proceedings of the Human Factors and Ergonomics Society 37th Annual Meeting*, 1993, pp. 30–33.
8. Ward, M., McDonald, N., An European Approach to the Integrated Management of Human Factors in Aircraft Maintenance: Introducing the IMMS, *Proceedings of the 7th International Engineering Psychology and Cognitive Ergonomics Conference*, 2007, pp. 852–859.
9. Robertson, M.M., Using Participatory Ergonomics to Design and Evaluate Human Factors Training Programs in Aviation Maintenance Operations Environments, *Proceedings of the XIVth Triennal Congress of the International Ergonomics Association and 44th Annual Meeting of the Human Factors and Ergonomics Association*, 2000, pp. 692–695.
10. Gagne, R., Briggs, L., Wagner, R., *Principles of Instructional Design*, Holt, Rinehart, and Winston, Inc., New York, 1988.
11. Goldstein, I. L., *Training in Organizations*, Wadsworth Publishing, Belmont, CA, 1993.
12. Knirk, F.G., Gustafson, K.I., *Instructional Technology: A Systematic Approach to Education*, Holt, Reinhart, and Winston, New York, 1986.
13. Kirkpatrick, D., Techniques for Evaluating Training Programs, *Training and Development Journal*, Vol. 31, No. 11, 1979, pp. 9–12.
14. Gordon, S., *Systematic Training Program Design: Maximizing and Minimizing Liability*, Prentice Hall, Englewood Cliffs, NJ, 1994.
15. Hannum, W., Hansen, C., *Instructional Systems Development in Large Organizations*, Prentice Hall, Inc., Englewood Cliffs, NJ, 1992.
16. Johnson, W.B., Human Factors Training for Aviation Maintenance Personnel, *Proceedings of the Human Factors and Ergonomics Society 41st Annual Meeting*, 1997, pp. 1168–1171.
17. Johnson, W.B., The National Plan for Aviation Human Factors: Maintenance Research Issues, *Proceedings of the Human Factors Society 35th Annual Meeting*, 1991, pp. 28–32.

7 Human Factors in Power Plant Maintenance

7.1 INTRODUCTION

Human factors play an important role in power plant maintenance because improving the maintainability design of power plant facilities, systems, and equipment with respect to human factors helps to increase, directly or indirectly, plant productivity, availability, and safety. For example, past experiences indicate that many plant outages have been either caused or prolonged by human factors problems associated with maintenance. It is estimated that the loss of plant power generation costs at least $500,000 to $750,000 per day [1].

Interest in human factors issues in the power industry is relatively new in comparison to the aerospace industry. In fact, it may be traced back to the middle of the 1970s when the WASH-1400 Reactor Safety Study criticized the deviation of the design of controls and displays and their arrangement in nuclear power plants from the human factors engineering standards [2]. The Electric Power Research Institute (EPRI) took note of this criticism and sponsored a study concerning the review of human factors in nuclear power plant control rooms in the United States [3]. This study highlighted various minor and major human factors–related deficiencies that can result in the poor effectiveness of the man-machine interface [1, 3]. Subsequently, over the years, the occurrence of many human factors–deficiency-related events, including the Three Mile Island nuclear power plant accident, has resulted in an increased attention to human factors in various areas of power generation including maintenance.

This chapter presents various important aspects of human factors in power plant maintenance.

7.2 HUMAN FACTORS ENGINEERING MAINTENANCE– RELATED DEFICIENCIES IN POWER PLANT SYSTEMS

Over the years, many studies have identified various human factors–engineering deficiencies, directly or indirectly maintenance related, in power plant systems. One survey-based study has classified such deficiencies under six categories [4]. In descending order, these categories are as follows [4]:

Limited access or inadequate clearance to perform maintenance. It means that there is inadequate clearance for inspection, no room for the right tool, etc.

Equipment poorly designed to facilitate the maintenance activity effectively. It means that the required work is too detailed to perform with mask and gloves on, design is complicated (i.e., too difficult to repair), it is impossible to open cabinet doors all the way, etc.

Equipment/systems inherently unreliable. It means items such as cheap and dirty design of the rod position indicators, flatbed fitter is under-designed and requires constant maintenance, overly sensitive controllers, and the system drifts and is unstable.

Personnel safety hazard. It means items such as no safety rail where is, say, a 35-foot drop, oil on the floor from the main feed pumps, hydrogen unloading facility is rather dangerous, and poorly designed equipment in high-radiation areas.

Impaired mobility for both personnel and equipment. It means items such as no elevator access to the turbine deck, lack of work platforms with ladders, one way access to hatch into containment, no cargo elevators where needed, and lack of pad eyes for lifting.

Miscellaneous. It means items such as lack of standardization, high-temperature environment, and poor air conditioning.

7.3 DESIRABLE HUMAN FACTORS ENGINEERING MAINTENANCE–RELATED ATTRIBUTES OF WELL-DESIGNED SYSTEMS IN POWER GENERATION

The survey-based study of Ref. [4] reported many desirable human factors–, engineering maintenance–related attributes of well-designed systems used in power generation. In descending order, these attributes are as follows [4]:

- **Effective accessibility.** It means items such as good accessibility around the diesels, easy access to air compressors, and good access to rod controls for repair.
- **Ease of disassembly, removal, and repair.** It means items such as modular design of rod controls, easy removal of circuit breakers, and modules on rollout rails.
- **Ease of system troubleshooting testing, and monitoring.** It means items such as engineered guards easy to test, built-in calibration system, good test jacks and easy to input signals, and control cabinet for boiler control easy to troubleshoot.
- **Effective lifting and movement capability.** It means items such as built-in hoist always in place, easy removal through roof, and access for vehicles.
- **Highly reliable equipment.** It means items such as reliable relays, air compressor easy to operate and rarely breaks down, and highly reliable engineered safeguards actuation system.
- **Ease of inspection and servicing.** It means items such as good access for preventive maintenance, easy to spot problems, and ease of oil changes.
- **Good quality prints and manuals.** It means items such as readable prints, understandable procedures, and detailed operating instructions.

- **Avoidance of contaminated areas.** It basically means, for example, equipment placed in an accessible location well outside the "hot" areas.
- **Good laydown area.** It means, for example, excellent laydown area for turbine-generator.
- **Availability of required tools.** It means the availability of all necessary tools, for example, all the essential special tools provided for a complicated assembly.
- **Miscellaneous.** It includes items such as fail-safe design and frequent use of mock-ups for training.

7.4 POWER GENERATION PLANT PERFORMANCE GOALS THAT DRIVE DECISIONS ABOUT HUMAN FACTORS

There are many power generation plant performance goals that drive, directly or indirectly, maintenance-related decisions about human factors. These goals may be grouped under three classifications as shown in Figure 7.1 [5]. The classifications are plant safety, plant productivity, and plant availability. The plant safety goals include minimizing injury to personnel, damage to equipment, and, in the case of nuclear power plants, eliminating the potential for release of radioactivity to the environment and reducing the radiation exposure to humans.

The plant availability goals include increasing the amount of time the plant can operate at full power generation capacity by minimizing the occurrence of human errors that, directly or indirectly, contribute to system/equipment failures or increase system/equipment corrective maintenance time.

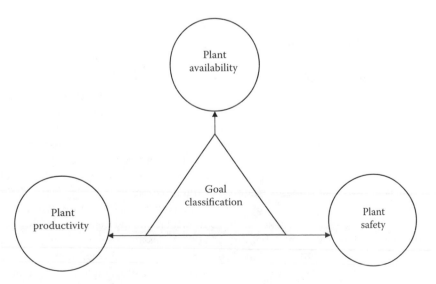

FIGURE 7.1 Classifications of power generation plant performance goals that drive decisions about human factors.

Finally, the plant productivity goals include improving the reliability, efficiency, and motivation of all involved personnel.

7.5 STUDY OF HUMAN FACTORS IN POWER PLANTS

A survey of the maintenance of five nuclear and four fossil-fuel power generation plants with respect to human factors revealed various types of, directly or indirectly, human factors–related problems. This study was wide ranging in scope, extending to an examination of items such as facilities, environmental factors, designs, organizational factors, procedures, spares, and tools.

The study findings were grouped under the following 16 classifications [1, 4]:

- Facility design factors
- Environmental factors
- Equipment maintainability
- Anthropometrics and human strength
- Movement of humans and machines
- Labeling and coding
- Maintenance stores, supplies, and tools
- Maintenance information, procedures, and manuals
- Personnel safety
- Communications
- Equipment protection
- Productivity and organizational interfaces
- Preventive maintenance and malfunction diagnosis
- Job practices
- Selection and training
- Maintenance errors and accidents

Some of the above classifications are described below [1, 4].

The facility design factors classification is concerned with, directly or indirectly, human factors–related problems pertaining to facility design. Some of these problems are high noise levels, poor temperature-ventilation control, inadequate facility to store contaminated equipment, and insufficient storage space to satisfy maintenance needs effectively. The environmental factors classification is concerned with human factors problems pertaining to the environment. Two examples of these problems are heat stress and a high variability of illumination.

An example of the problems belonging to the anthrometrics and human strength classification is the lack of easy access to equipment requiring maintenance. Some of the problems belonging to the labeling and coding classification are poorly descriptive label tags, under-estimation of the need for identifying information by designers, unsystematic replacement of labels lost or obscured over time, high likelihood of the occurrence of maintenance errors in multi-unit plants in which both units are identical or highly similar in appearance.

Some of the problems belonging to the personnel safety classification are radiation exposure, steam burns, chemical burns, and heat prostration. The communications classification includes problems such as inadequate capacity of the existing communications system to satisfy the volume of communications traffic required throughout the plant, particularly during outages, the protective clothing worn by maintenance personnel while working in radioactive environment causes serious impediments to effective communications and insufficient communication coverage throughout the plant.

The most common problem belonging to the equipment maintainability classification is the placement of equipment parts in locations that are inaccessible from a normal work position. The main problem belonging to the maintenance information, procedures, and manuals classification is poorly written procedures and inadequate manuals. Two problems belonging to the selection and training classification are the informality of the training process, with no clearly defined selection criteria and lacking validated screening tools or techniques and the overall inadequacy of the training efforts in nuclear power plants.

7.6 HUMAN FACTORS APPROACHES FOR ASSESSING AND IMPROVING POWER PLANT MAINTAINABILITY

Many human factors methods can be used to assess and improve power plant maintainability. Six of these methods are shown in Figure 7.2 [1, 6]. Each of these methods is described below, separately.

7.6.1 TASK ANALYSIS

This is a systematic approach used to assess the equipment maintainer's needs for successfully working with hardware to accomplish a given task. The analyst records and oversees each task element and start and completion times, in addition to making observations concerning impediments to effective maintainability. The observations

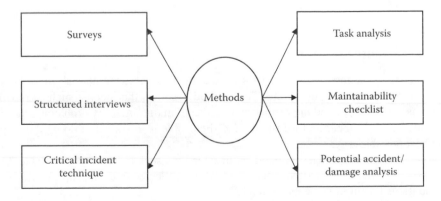

FIGURE 7.2 Human factors methods for assessing and improving power plant maintainability.

are categorized under sixteen classifications: equipment maintainability design features, availability of appropriate maintenance information (e.g., schematics, procedures, and manuals), tools and job aids, maintenance crew interactions, equipment damage potential, decision-making factors, personnel hazards, lifting or movement aids, communication, training needs, spare-parts retrieval, workshop adequacy, supervisor-subordinate relationships, environmental factors, access factors, and facility design features [1].

Additional information on the method is available in Refs. [1, 6].

7.6.2 MAINTAINABILITY CHECKLIST

This checklist is based primarily on the survey study reported in Ref. [4] and is divided into 14 distinct topical areas. These areas are personnel safety, radiation protection, communications, facilities, maintenance information, equipment maintainability, labeling and coding, preventive maintenance, anthropometrics and human strength, selection and training, environmental factors, job and organizational factors, equipment protection, and stores, spares, and tools.

Additional information on the method is available in Ref. [4].

7.6.3 POTENTIAL ACCIDENT/DAMAGE ANALYSES

This is a structured approach used to assess the accident, damage, or potential error inherent in a specified task. To determine the potential for the occurrence of mishaps in the performance of a maintenance job, the starting point is to establish a mechanism that describes the job under consideration in detail. Subsequently for each task element the following question is asked by the interviewer of the interviewee (e.g., repair person): Is there a low, medium, or high potential for the occurrence of an error/an accident/damage to equipment /system in performing, say, step xyz?

After a careful analysis of all the collected data, changes to items such as equipment, facility, and procedures are recommended. Additional information on the method is available in Refs. [1, 6].

7.6.4 STRUCTURED INTERVIEWS

This is one of the most effective methods to collect valuable maintainability-related data in the shortest possible time. The method assumes that people such as repair persons, technicians, and their supervisors close to maintainability problems usually provide the most meaningful insights into the problems involved in doing their job the best possible way.

In a structured interview, a fixed set of questions such as presented in Table 7.1, are asked [1, 6]. After a careful analysis of all the collected data, appropriate recommendations for improvements are made.

Additional information on the method is available in Refs. [1, 6].

TABLE 7.1
A Sample of Questions Asked during a Structured Interview

No.	Questions
1.	Is your workshop facility arranged properly so that it allows efficient and safe performance of maintenance activities?
2.	How well is your workshop facility integrated into the total plant design?
3.	How would you describe the environment in your workshop facility with respect to factors such as ventilation, illumination, and noise?
4.	Are proper laydown areas and workbenches provided?
5.	Is our workshop facility sized appropriately to accommodate effectively all the personnel in your organization?

7.6.5 CRITICAL INCIDENT TECHNIQUE

Past experiences indicate that the history of maintenance errors, accidents, or near-mishaps can provide information concerning required maintainability improvements. The critical incident technique is an effective tool to examine such case histories from the human-factors standpoint. The application of the critical incident technique calls for making arrangements to meet individually with members of the maintenance organization. The following three questions are asked of each individual:

- Give one example of a maintenance error, accident, or near mishap with serious or potentially serious consequences, based on your personal experience. In addition, describe the specifics of the case involved and indicate the ways the situation could have been averted.
- Give one example of a plant system or unit of equipment that is not "human engineered" or is poorly designed from the maintenance person's perspective and which has resulted in, or could result in a safety hazard, damage to equipment, or an error.
- Give one example of a plant system or unit of equipment that is well– "human–engineered" or quite straightforward to maintain, and describe the system/unit by emphasizing the features that make it good from the maintainer's perspective.

After the analysis of all the collected data, appropriate changes for improvements are recommended.

Additional information on the method is available in Refs. [1, 6].

7.6.6 SURVEYS

When the results obtained through the application of methods such as task analysis, structured interviews, and maintainability checklist, indicate a need for more detailed examination of certain maintainability-related factors; the surveys method is used. Two examples of such scenario are as follows:

- Poor illumination is proving to be a problem in the course of analyzing one or more specific tasks. Under such condition, it might be useful to conduct a plant-wide illumination survey of all maintenance work-related sites.
- The majority of maintenance manpower has expressed concerns in the area of communications. Under such conditions, it might be useful to conduct a survey or test of message intelligibility between important communication links within the plant.

Additional information on conducting such surveys is available in Ref. [6].

7.7 BENEFITS OF HUMAN FACTORS ENGINEERING APPLICATIONS IN POWER GENERATION

Past experiences indicate that there are many benefits of human factors–engineering applications in power plant maintenance. Nonetheless, benefits of human factors engineering applications in power generation in general that directly or indirectly concern maintenance may be grouped under two main categories: increments and reductions [5, 7–9]. The increments category includes increases in safety, productivity, and availability, reliability and efficiency of personnel performance, adequacy of communications, cost-effectiveness of training, and job satisfaction of personnel (i.e., motivation, confidence, and commitment to achieving plant goals).

The reductions category includes reduction in needless costs, occurrence of human error, consequences of error (i.e., number and severity of injuries and damage to equipment), wasted time and motion, number and qualifications of personnel required, training requirements and attrition, and job dissatisfaction of personnel (i.e., turnover and absenteeism).

7.8 PROBLEMS

1. Write an essay on human factors in power plant maintenance.
2. List and discuss at least five types of human factors engineering maintenance–related deficiencies in power plant systems.
3. List and discuss at least eight desirable human factors engineering maintenance–related attributes of well-designed systems in power generation.
4. What are the power generation plant performance goals that drive, directly or indirectly, maintenance-related decisions about human factors?
5. What are the important benefits of human factors–engineering application in power generation with respect to the maintenance activity?
6. What were the classifications of the findings in a survey of maintenance of nuclear and fossil-fuel power generation plants with respect to human factors?
7. List at least five human factors methods that can be used to assess and improve power plant maintainability.

8. Discuss the following two human factors approaches that can be used to assess and improve power plant maintainability:
   • Task analysis
   • Maintainability checklist
9. Describe the structured interviews method that can be used to evaluate and enhance power plant maintainability and give a sample of questions asked during a structured interview.
10. Compare the critical incident technique with the potential accident/damage analyses method.

REFERENCES

1. Seminara, J.L., Parsons, S.O., Human Factors Engineering and Power Plant Maintenance, *Maintenance Management International*, Vol. 6, 1985, pp. 33–71.
2. WASH-1400, Reactor Safety Study: An Assessment of Accident Risks in U.S. Commercial Nuclear Power Plants, U.S. Nuclear Regulatory Commission, Washington, D.C., 1975.
3. Seminara, J.L., Gonzalez, W.R., Parsons, S.O., Human Factors Review of Nuclear Power Plant Control Room Design, Report No. EPRI NP-309, Electric Power Research Institute (EPRI), Palo Alto, CA, 1976.
4. Seminara, J.L., Parsons, S.O., Human Factors Review of Power Plant Maintainability, Report No. EPRI NP-1567, Electric Power Research Institute (EPRI), Palo Alto, CA, 1981.
5. Kinkade, R.G., Human Factors Primer for Nuclear Utility Managers, Report No. EPRI NP-5714, Electric Power Research Institute (EPRI), Palo Alto, CA, 1988.
6. Seminara, J.L., Human Factors Methods for Assessing and Enhancing Power Plant Maintainability, Report No. EPRI NP-2360, Electric Power Research Institute, Palo Alto, CA, 1982.
7. Annual Report, Electric Power Research Institute (EPRI), Palo Alto, California, 1982.
8. Parfitt, B., First Use: Frozen Water Garment Use at TMI-2, Report No. EPRI 4102B (RP 1705), Electric Power Research Institute (EPRI), Palo Alto, CA, 1986.
9. Shriver, E.L., Zach, S.E., Foley, J.P., Test of Job Performance Aids for Power Plants, Report No. EPRI NP-2676, Electric Power Research Institute, Palo Alto, CA, 1982.

8 Human Error in Aviation Maintenance

8.1 INTRODUCTION

Maintenance is an important element of the aviation industry worldwide, and in 1989 U.S. airlines spent around 12% of their operating costs on the maintenance activity [1, 2]. During the period from 1980 to 1988, the cost of airline maintenance increased from about $2.9 billion to $5.7 billion [3]. This increase is attributable to factors such as increase in air traffic and increased maintenance for continuing airworthiness of aging aircraft.

Needless to say, increase in air traffic and increased demands on aircraft utilization because of the stringent requirements of commercial schedules continue to put significant pressures on the maintenance activity for on-time performance. In turn, this has increased chances for the occurrence of human errors in aircraft maintenance operations [4]. A study conducted in the United Kingdom reported that the occurrence of maintenance error events per million flights has doubled during the period from 1990 to 2000 [5]. This clearly indicates that there is a need to eliminate or minimize the occurrence of such error events for reliable and safe flights.

This chapter presents various importance aspects of human error in aviation maintenance.

8.2 FACTS, FIGURES, AND EXAMPLES

Some of the facts, figures, and examples directly or indirectly concerned with the occurrence of human error in aviation maintenance are as follows:

- A study revealed that approximately 18% of all aircraft accidents are maintenance related [6, 7].
- As per Ref. [8] maintenance error contributes to 15% of air carrier accidents and costs the United States industry over $1 billion dollars annually.
- According to a Boeing study 19.1% of in-flight engine shutdowns are caused by maintenance error [8].
- A study reported that maintenance and inspection are the factor in approximately 12% of major aircraft accidents [9, 10].
- A study of 122 maintenance errors occurring in a major airline over a period of three years revealed that their breakdowns were: omission (56%), wrong installations (30%), incorrect parts (8%), and other (6%) [11, 12].

- An analysis of safety issues versus onboard fatalities among jet fleets world-wide during the period 1982–1991 identified maintenance and inspection as the second most important safety issue with onboard fatalities [13, 14].
- In 1979, 272 people were killed in a DC-10 aircraft accident due to improper maintenance procedures followed by maintenance personnel [15].
- In 1991, 13 people were killed in an Embraer 120 aircraft accident due to a human error during scheduled maintenance [4, 5].
- In 1988, the upper cabin structure of a Boeing 737-200 aircraft was ripped away during a flight because of structural failure, basically due to the failure of maintenance inspectors to identify over 240 cracks in the aircraft skin during the inspection process [5, 16].

8.3 CAUSES OF HUMAN ERROR IN AVIATION MAINTENANCE AND MAJOR CATEGORIES OF HUMAN ERRORS IN AVIATION MAINTENANCE AND INSPECTION TASKS

There are many factors that can impact performance of aviation maintenance personnel. Over 300 such factors/influences are listed in a document prepared by the International Civil Aviation Organization [17]. These factors/influences range from boredom to temperature. Some of the important reasons, directly or indirectly, for the occurrence of human error in aviation maintenance are time pressure; inadequate training, work tools, and experience; complex maintenance tasks, poorly written maintenance procedures, poor equipment design, outdated maintenance manuals, poor work layout, fatigued maintenance personnel, and poor work environment (e.g., temperature, humidity, lighting) [15,18].

There are many major categories of human errors in aviation maintenance and inspection-related tasks. Eight of these categories are incorrect assembly sequence (e.g., incorrect sequence of inner cylinder spacer and lock ring assembly), procedural defects (e.g., nose landing gear door not closed), wrong part (e.g., incorrect pitot-static probes installed), incorrect configuration (e.g., valve inserted in backward direction), missing part (e.g., bolt-nut not secured), defective part (e.g., worn cables, fluid leakage, cracked pylon, etc.), functional defects (e.g., wrong tire pressure), and tactile defects (e.g., seat not locking in correct position) [12, 19, 20].

8.4 TYPES OF HUMAN ERROR IN AIRCRAFT MAINTENANCE AND THEIR FREQUENCY

In 1994, a Boeing study examined a total of 86 aircraft incident reports with respect to maintenance error and reported 31 types of maintenance errors. These types, along with their frequency in parentheses, are: system operated in unsafe conditions (16), system not made safe (10), equipment failure (10), towing event (10), falls and spontaneous actions (6), degradation not discovered (6), person entered dangerous zones (5), unfinished installation (5), work not documented (5), did not obtain or use appropriate equipment (4), person contacted hazard (4), unserviceable equipment used (4), equipment not activated/deactivated (4), no appropriate verbal warning given (3), safety lock or warning moved (2), pin/tie left in place (2), not tested appropriately (2), equipment/vehicle contacted aircraft (2), warning sign or tag not used (2), vehicle driving instead of towing (2), wrong fluid type (1), access panel not closed (1), wrong panel installation (1), material left in engine/aircraft (1), incorrect orientation (1), equipment not installed (1), contamination of open system (1), wrong component/equipment installed (1), unable to access part or component in stores (1), necessary servicing not performed (1), and miscellaneous (6) [21].

8.5 COMMON HUMAN ERRORS IN AIRCRAFT MAINTENANCE ACTIVITIES

Over the years various studies have identified commonly occurring human errors in aircraft maintenance activities. One of these studies conducted by the United Kingdom Civilian Aviation Authority (UKCAA) over a period of three years has identified a total of eight commonly occurring human errors in aircraft maintenance, as shown in Figure 8.1 [12, 22].

8.6 AIRCRAFT MAINTENANCE ERROR ANALYSIS METHODS

Over the years, many methods have been developed in reliability and its associated areas that can be used to perform human error analysis in the area of aircraft maintenance. Three of these methods are presented below.

8.6.1 CAUSE-AND-EFFECT DIAGRAM

This diagram was developed by a Japanese man named K. Ishikawa in the early 1950s. It is also referred to in the published literature as an Ishikawa diagram or a

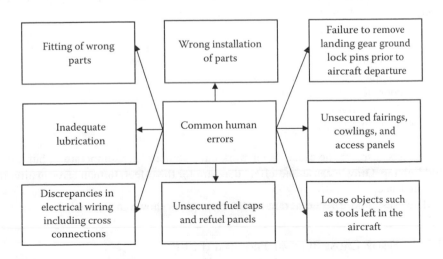

FIGURE 8.1 Commonly occurring human errors in aircraft maintenance.

fishbone diagram. The diagram can be a useful tool to determine the root causes of a specified aircraft maintenance error and generate appropriate relevant ideas.

Pictorially, the box on the extreme right-hand side of the diagram represents effect and the left-hand side represents all the possible causes that are connected to the centerline. In turn, usually each cause is composed of various subcauses. Usually, the following five steps are followed to develop a cause-and-effect diagram [8]:

- **Step 1:** Develop problem statement.
- **Step 2:** Brainstorm to identify possible causes.
- **Step 3:** Establish major cause categories by stratifying into natural groupings and process steps.
- **Step 4:** Develop the diagram by connecting all the causes by following the appropriate process steps and fill in the effect (i.e., the problem) in box on the right hand side of the diagram.
- **Step 5:** Refine cause categories/classifications by asking questions such as what causes this? And why does this condition exist?

There are many benefits of the cause-and-effect diagram. Some of the important ones are as follows:

- An effective tool for generating ideas
- An effective approach to present an orderly arrangement of theories
- A useful tool to identify root causes
- A useful approach for guiding further inquiry

EXAMPLE 8.1

A study of aircraft maintenance facility reported the following six causes for the occurrence of human error in maintenance:

- Poor work environment
- Time pressure
- Complex maintenance tasks
- Poor equipment design
- Poor work layout
- Inadequate tools

Three subcauses of the cause "poor work environment" are temperature, humidity, and lighting. Draw a cause-and-effect diagram for the effect: human error in aircraft maintenance.

The cause-and-effect diagram for the example is shown in Figure 8.2.

8.6.2 ERROR-CAUSE REMOVAL PROGRAM (ECRP)

This method was originally developed to reduce the occurrence of human error to some tolerable level in production operations [24]. It can also be used to reduce

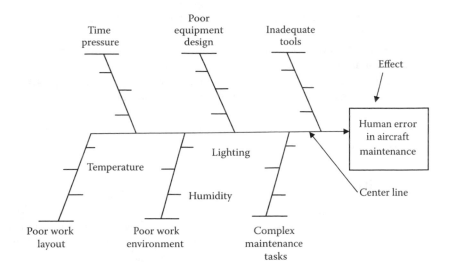

FIGURE 8.2 Cause-and-effect diagram for the occurrence of human error in aircraft maintenance.

human error in aircraft maintenance operations. The emphasis of this method is on preventive measures rather than merely on remedial ones. In terms of aircraft maintenance, ECRP may simply be described as the maintenance worker-participation program for reducing the occurrence of human errors.

More specifically, the ECRP is composed of teams of workers (e.g., aircraft maintenance workers) with each team having its own coordinator, who has special technical and group-related skills. Workers present their error and error-likely reports during team meetings held periodically. After appropriate discussions on these reports, recommendations are made for preventive or remedial measures. Team coordinators present the recommendations to management for appropriate actions.

The seven basic elements of the ECRP are as follows [24]:

- All people involved with the ECRP are educated about the usefulness of the ECRP.
- All maintenance workers and team coordinators are trained in data collection and analysis approaches.
- The efforts of the aircraft maintenance workers in regard to ECRP are recognized appropriately by the management.
- Human factors and other specialists determine the effects of changes made in, say, aircraft maintenance operations with the aid of the ECRP inputs.
- The most promising proposed solutions are fully implemented by the management.
- All proposed solutions are evaluated with respect to cost by various specialists including human factors specialists.
- Aircraft maintenance workers report and evaluate errors and error-likely conditions, in addition to proposing solutions to eradicate error causes.

Finally, three of the important guidelines concerning ECRP are as follows:

- Focus on data collection on items such as error-likely conditions, accident-prone conditions, and errors.
- Evaluate each work redesign recommended by the team with respect to factors such as increments in cost-effectiveness and job satisfaction, and the degree of error redaction.
- Restrict to the identification of work situations that require redesign for reducing the error occurrence potential.

8.6.3 FAULT TREE ANALYSIS

This is a powerful and flexible method often used in industry to perform various types of reliability-related analysis. The method is described in Chapter 4 and in Refs. [18, 20]. Its application to perform human error analysis in aviation maintenance is demonstrated through the example presented below.

EXAMPLE 8.2

Assume that the subcauses of the cause "poor work environment" in Example 8.1 are poor lighting, high/low temperature, and distractions. Similarly, the subcauses of the cause "poor equipment design" are poorly written design specification, no formal consideration given to the occurrence of maintenance error in design specification, and misinterpretation of design specification.

Develop a fault tree for Example 8.1, for top event "Human error in aircraft maintenance" by considering the above subcauses and using fault tree symbols given in Chapter 4.

A fault tree for the example is shown in Figure 8.3.

EXAMPLE 8.3

Assume that the probability of occurrence of events in the circles (i.e., $X_1, X_2, X_3, \ldots, X_8$) shown in Figure 8.3 is 0.02. For independent events, calculate the probability of occurrence of the top event T (i.e., human error in aircraft maintenance), and intermediate events I_1, (i.e., poor equipment design) and I_2 (i.e., poor environment).

Using Chapter 4 and Refs. [18, 20], and the specified data values, we obtain the values of I_1, I_2, and T as follows:

The probability of occurrence of intermediate event I_1 is given by

$$P(I_1) = 1 - \{1 - P(X_1)\}\{1 - P(X_2)\}\{1 - P(X_3)\}$$
$$= 1 - \{1 - 0.02\}\{1 - 0.02\}\{1 - 0.02\}$$
$$= 0.0588$$

where $P(I_1)$, $P(X_1)$, $P(X_2)$, and $P(X_3)$ are the probabilities of occurrence of events I_1, X_1, X_2, and X_3, respectively.

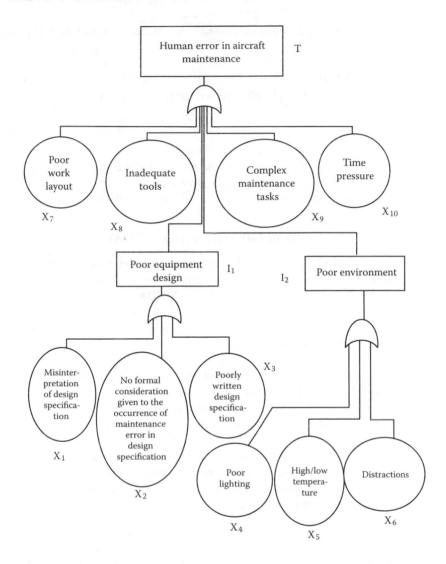

FIGURE 8.3 Fault tree for Example 8.2.

The probability of occurrence of intermediate event I_2 is expressed by

$$P(I_2) = 1 - \{1 - P(X_4)\}\{1 - P(X_5)\}\{1 - P(X_6)\}$$
$$= 1 - \{1 - 0.02\}\{1 - 0.02\}\{1 - 0.02\}$$
$$= 0.0588$$

where $P(I_2)$, $P(X_4)$, $P(X_5)$, and $P(X_6)$ are the probabilities of occurrence of events I_2, X_4, X_5, and X_6, respectively.

By using the above specified and calculated values, Chapter 4, and Refs. [18, 20] we obtain

$$P(T) = 1 - \{1 - P(X_7)\}\{1 - P(X_8)\}\{1 - P(X_9)\}\{1 - P(X_{10})\}\{1 - P(I_1)\}\{1 - P(I_2)\}$$
$$= 1 - \{1 - 0.02\}\{1 - 0.02\}\{1 - 0.02\}\{1 - 0.02\}\{1 - 0.0588\}\{1 - 0.0588\}$$
$$= 0.1829$$

where $P(T)$ is the probability of occurrence of event T.

Thus, the probabilities of occurrence of the top event T (i.e., human error in aircraft maintenance), intermediate event I_1 (i.e., poor equipment design), and intermediate event I_2 (i.e., poor environment) are 0.1829, 0.0588, and 0.0588, respectively.

8.7 MAINTENANCE ERROR DECISION AID (MEDA)

This important tool to investigate contributing factors to maintenance errors in aviation was developed by Boeing, along with industry partners such as Continental Airlines and United Airlines, in the 1990s [25–27]. MEDA may simply be described as a structured process for investigating the causes of human errors made by aircraft maintenance personnel. The philosophy of the process is shown in Figure 8.4 [26].

Four main objectives of the MEDA are as follows [27]:

- To highlight aircraft maintenance system-related problems that increase exposure to human error and decrease efficiency
- To provide the aircraft maintenance organization a better understanding of how human-performance-associated issues contribute to the occurrence of human error
- To provide the line-level aircraft maintenance personnel a standardized mechanism to investigate the occurrence of maintenance errors
- To provide an appropriate means of human error trend analysis for the aircraft maintenance organization

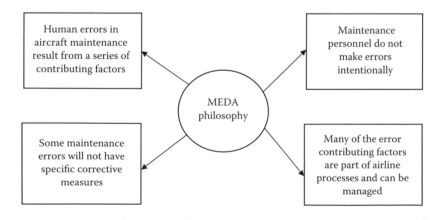

FIGURE 8.4 The philosophy of MEDA.

All in all, MEDA provides people associated with the aircraft maintenance activity a basic five-step process to follow: event, decision, investigation, prevention strategies, and feedback [25]. Additional information on these steps and on MEDA is available in Refs. [25–27].

8.8 USEFUL GUIDELINES FOR REDUCING HUMAN ERROR IN AIRCRAFT MAINTENANCE ACTIVITIES

Over the years, various guidelines have been developed for reducing human error in aircraft maintenance activities. These guidelines cover many areas as shown in Figure 8.5 [14, 20]. Two important guidelines concerning the area of design are as follows:

- Actively seek relevant information on human error occurrence during the maintenance phase, for providing effective inputs in the design phase.
- Ensure that equipment manufacturers give proper attention to maintenance-related human factors during the design phase.

Two guidelines in the area of tools and equipment are as follows:

- Review systems by which items such as lighting systems and stands are kept for removing unserviceable equipment from service and repairing it rapidly.
- Ensure the storage of all lockout devices in such a manner that it becomes immediately apparent when they are left in place inadvertently.

Some of the guidelines concerning risk management are to avoid performing simultaneously the same maintenance task on similar redundant units, review formally the effectiveness of defenses, such as engine runs, built into the system for detecting maintenance errors, and review the need to disturb normally operating systems to carry out rather nonessential periodic maintenance, because the disturbance may lead to a maintenance error.

A useful guideline in the area of communication is to ensure that proper systems are in place to disseminate important pieces of information to all individuals concerned with maintenance, so that repeated errors or changing procedures are considered with care.

Two particular guidelines in the area of training are as follows:

- Provide on a periodic basis training courses to all maintenance personnel with emphasis on company procedures.
- Consider introducing crew resourcement for personnel involved with the maintenance activity.

Some of the useful guidelines concerned with procedures are ensuring that standard work practices are being followed throughout aircraft maintenance operations,

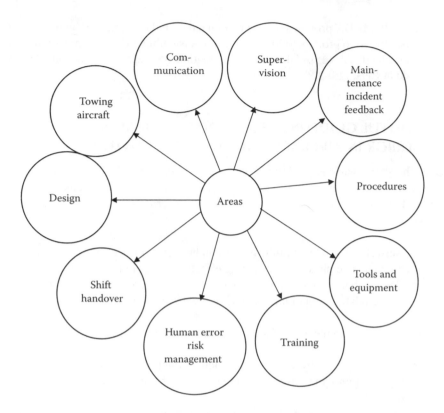

FIGURE 8.5 Areas covered by guidelines for reducing human error in aircraft maintenance activities.

reviewing maintenance work practices regularly to ensure that they do not vary significantly from formal procedures, and reviewing all documented maintenance procedures and practices periodically with respect to items such as accessibility, consistency, and realism.

A useful guideline in the area of supervision is to recognize that management and supervision-related oversights must be strengthened, particularly in the final hours of all shifts, as the occurrence of errors becomes more likely. Two particular guidelines pertaining to maintenance incident feedback are as follows:

- Ensure that all individuals associated with the training activity are provided proper feedback on the occurrence of human factors–related maintenance incidents regularly, so that appropriate corrective actions aimed at these problems are taken effectively.
- Ensure that all management personnel are given effective feedback on the occurrence of human factors–related maintenance incidents regularly, with proper consideration to the conditions that play an instrumental role in the occurrence of such incidents.

A guideline pertaining to the area of towing aircraft is to review the equipment and procedures used for towing to and from maintenance facilities on a regular basis. Finally, one particular guideline concerning shift handover is to ensure the effectiveness of practices associated with shift handover by considering factors such as communication and documentation, so that incomplete tasks are transferred correctly across all shifts.

8.9 CASE STUDIES IN HUMAN ERROR IN AVIATION MAINTENANCE

Over the years, many aircraft accidents directly or indirectly due to maintenance error have occurred throughout the world. Three such accidents are briefly described below.

8.9.1 CONTINENTAL EXPRESS EMBRAER 120 ACCIDENT

This accident occurred on September 11, 1991, when a Continental Express Embraer 120 aircraft crashed in Texas, killing all persons on board, because the leading edge of the left horizontal stabilizer separated from the aircraft [1, 20, 28]. An investigation into the accident reported that the night prior to the accident some maintenance work, involving the removal of a screw from the upper left surface of the "T-tail" of the aircraft, was carried out. When the shift change occurred, the maintenance work was only partially accomplished and it was not documented at all.

The maintenance personnel of the incoming shift, being totally unaware of the partial accomplishment of the maintenance work, signed the Embraer 120 back into service. The National Transportation Safety Board (NTSB), in its final report on the accident, identified poor maintenance practices within the airline organization [29]. Additional information on the accident is available in Ref. [29].

8.9.2 AIR MIDWEST RAYTHEON (BEECHCRAFT) 1900D ACCIDENT

This accident occurred on January 8, 2003, when an Air Midwest Raytheon 1900D aircraft lost pitch control during takeoff and crashed in North Carolina, killing all persons on board (19 passengers and 2 crew members). Some of the factors that contributed to the cause of the accident were as follows [28]:

- The contractor's quality-assurance inspector's total failure to detect the wrong rigging of the elevator control system
- The operator's maintenance procedures and documentation and lack of oversight of the work being performed at the maintenance station
- The regulator's lack of oversight of the maintenance program of the operator

Additional information on the accident is available in Ref. [28].

8.9.3 British Airways BAC1-11 Accident

This accident occurred on June 10, 1990, when a British Airways BAC1-11 aircraft departed from Birmingham Airport in the United Kingdom to a destination in Spain, carrying 81 passengers and 6 crew members. During the aircraft's climb through 17,300 feet altitude, a cockpit windscreen was blown out. Consequently, the pilot in command was sucked out through the windscreen aperture [4].

The copilot immediately regained control of the aircraft and the other crew members held the pilot by the ankles until the safe landing of the aircraft. A subsequent investigation into the accident reported that the cause of the accident was the fitting of a replacement windscreen by maintenance workers using wrong bolts [4].

Additional information on the accident is available in Ref. [4].

8.10   PROBLEMS

1. Write an essay on human error in aviation maintenance.
2. List at least five facts and figures on human error in aviation maintenance.
3. What are the important reasons, directly or indirectly, for the occurrence of human error in aviation maintenance?
4. Discuss major categories of human errors in aviation maintenance and inspection tasks.
5. What are the commonly occurring human errors in aircraft maintenance?
6. Describe the error-cause removal program.
7. Describe the maintenance error decision aid (MEDA).
8. What are the useful guidelines to reduce human errors in the following areas of aircraft maintenance?
   • Tools and equipment
   • Shift-handover
   • Communication
9. Discuss two case studies concerned with human error in aviation maintenance.
10. Discuss the benefits of the cause-and-effect diagram.

REFERENCES

1. Hobbs, A., Williamson, A., Human Factors in Airline Maintenance, *Proceedings of the Conference on Applied Psychology*, 1995, pp. 384–393.
2. Shepherd, W.T., The FAA Human Factors Program in Aircraft Maintenance and Inspection, *Proceedings of the 5th Federal Aviation Administration (FAA) Meeting on Human Factors Issues in Aircraft Maintenance and Inspection*, June 1991, pp. 1–5.
3. Shepherd, W.T., Johnson, W.B., Drury, C.G., Berninger, D., Human Factors in Aviation Maintenance Phase One: Progress Report, Report No. AM-91/16, Office of Aviation Medicine, Federal Aviation Administration (FAA), Washington, D.C., November 1991.
4. Report No. CAP 718, Human Factors in Aircraft Maintenance and Inspection, Prepared by the Safety Regulation Group, Civil Aviation Authority, London, UK, 2002. Available from the Stationery Office, P.O. Box 29, Norwich, UK.

5. Report No. DOC 9824-AN/450, Human Factors Guidelines for Aircraft Maintenance Manual, International Civil Aviation Organization (ICAO), Montreal, Canada, 2003.
6. Kraus, D.C., Gramopadhye, A.K., Effect of Team Training on Aircraft Maintenance Technicians: Computer-Based Training Versus Instructor-Based Training, *International Journal of Industrial Ergonomics*, Vol. 27, 2001, pp. 141–157.
7. Phillips, E.H., Focus on Accident Prevention Key to Future Airline Safety, *Aviation Week and Space Technology*, 1994, Issue No. 5, pp. 52–53.
8. Marx, D.A., Learning from Our Mistakes: A Review of Maintenance Error Investigation and Analysis Systems (with Recommendations to the FAA), Federal Aviation Administration (FAA), Washington, D.C., January, 1998.
9. Marx, D.A., Graeber, R.C., Human Error in Maintenance, in *Aviation Psychology in Practice*, edited by N. Johnston, N. McDonald, and R. Fuller, Ashgate Publishing, London, 1994, pp. 87–104.
10. Gray, N., Maintenance Error Management in the ADF, *Touchdown* (*Royal Australian Navy*), December 2004, pp. 1–4. Also available online at http://www.navy.gov.au/publications/touchdown/dec.04/mainterr.html.
11. Graeber, R.C., Max, D.A., Reducing Human Error in Aircraft Maintenance Operations, *Proceedings of the 46th Annual International Safety Seminar*, 1993, pp. 147–160.
12. Latorella, K.A., Prabhu, P.V., A Review of Human Error in Aviation Maintenance and Inspection, *International Journal of Industrial Ergonomics*, Vol. 26, 2000, pp. 133–161.
13. Russell, P.D., Management Strategies for Accident Prevention, *Air Asia*, Vol. 6, 1994, pp. 31–41.
14. Report No. 2–97, Human Factors in Airline Maintenance: A Study of Incident Reports, Bureau of Air Safety Investigation (BASI), Department of Transport and Regional Development, Canberra, Australia, 1997.
15. Christensen, J.M., Howard, J.M., Field Experience in Maintenance, in *Human Detection and Diagnosis of System Failures*, edited by J. Rasmussen and W.B. Rouse, Plenum Press, New York, 1981, pp. 111–133.
16. Wenner, C.A., Drury, C.G., Analyzing Human Error in Aircraft Ground Damage Incidents, *International Journal of Industrial Ergonomics*, Vol. 26, 2000, pp. 177–199.
17. Report No. 93–1, Investigation of Human Factors in Accidents and Incidents, International Civil Aviation Organization, Montreal, Canada, 1993.
18. Dhillon, B.S., *Human Reliability: With Human Factors*, Pergamon Press, New York, 1986.
19. Prabhu, P., Drury, C.G., A Framework for the Design of the Aircraft Inspection Information Environment, *Proceedings of the 7th FAA Meeting on Human Factors Issues in Aircraft Maintenance and Inspection*, 1992, pp. 54–60.
20. Dhillon, B.S., *Human Reliability and Error in Transportation Systems*, Springer-Verlag, London, 2007.
21. Maintenance Error Decision Aid (MEDA), Developed by Boeing Commercial Airplane Group, Seattle, Washington, 1994.
22. Allen, J.P., Rankin, W.L., A Summary of the Use and Impact of the Maintenance Error Decision Aid (MEDA) on the Commercial Aviation Industry, *Proceedings of the 48th Annual International Air Safety Seminar*, 1995, pp. 359–369.
23. Besterfield, B.S., *Quality Control*, Prentice Hall, Upper Saddle River, NJ, 2001.
24. Swain, A.D., An Error-Cause Removal Program for Industry, *Human Factors*, Vol. 12, 1973, pp. 207–221.
25. Rankin, W., MEDA Investigation Process, *Aero Quarterly* (Boeing.com/commercial/aero magazine), Vol. 2, No. 1, 2007, pp. 15–22.

26. Rankin, W.L. Allen, J.P., Sargent, R.A., Maintenance Error Decision Aid: Progress Report, *Proceedings of the 11th FAA/AAM Meeting on Human Factors in Aviation Maintenance and Inspection*, 1997, pp. 19–24.

27. Hibit, R., Marx, D.A., Reducing Human Error in Aircraft Maintenance Operations with the Maintenance Error Decision Aid (MEDA), *Proceedings of the Human Factors and Ergonomics Society 38th Annual Meeting*, 1994, pp. 111–114.

28. Kanki, B.G., Managing Procedural Error in Maintenance, *Proceedings of the International Air Safety Seminar (IASS)*, 2005, pp. 233–244.

29. Report No. 92/04, Aircraft Accident Report on Continental Express, Embraer 120, National Transportation Safety Board (NTSB), Washington, D.C., 1992.

9 Human Error in Power Plant Maintenance

9.1 INTRODUCTION

Maintenance is an essential activity in power plants, and it consumes a significant amount of money spent on power generation. Human error in maintenance has been found to be an important factor in the causation of power generation safety-related incidents [1]. A study of reliability problem-related events concerning electrical/electronic components in nuclear power plants revealed that human errors made by maintenance personnel and technicians exceeded operator errors and that over three-quarters of the errors took place during the testing and maintenance activity [1, 2]. Furthermore, according to Refs. [1, 3], errors made during testing and maintenance caused reactor core melt more easily than did errors during operation.

The cost of maintenance errors, including restoration costs and opportunity costs, is potentially very high, the damage impact on the equipment may decrease its life quite considerably, and serious potential hazards to human lives may result. Because of potentially critical consequences such as these to system function and public safety, the prevention of human errors in maintenance tasks in power generation is receiving increasing attention.

This chapter presents various important aspects of human error in power plant maintenance.

9.2 FACTS AND FIGURES

Some of the facts, figures, and examples directly or indirectly related to human error in power plant maintenance are as follows:

- A study reported that over 20% of all system failures in fossil power plants occur due to human errors and maintenance errors account for about 60% of the annual power loss due to human errors [4].
- A number of studies reported that between 55% and 65% of human performance problems surveyed in power generation were associated with maintenance-related activities [5, 6].
- A study of over 4400 maintenance history records covering the period from 1992 to 1994, concerning a boiling water reactor (BWR) nuclear power plant, reported that around 7.5% of all failure records could be classified as human errors related to maintenance actions [7, 8].
- A study of 199 human errors that occurred in Japanese nuclear power plants from 1965 to 1995 revealed that around 50 of them were related to maintenance activities [9].

- A study of 126 human error-related significant events in 1990, in nuclear power generation, reported that 42% of the problems were linked to maintenance and modification [5].
- On Christmas Day in 1989, two nuclear reactors were shut down due to maintenance error and caused rolling blackouts in the state of Florida [10].
- A blast at the Ford Rouge power plant in Dearborn, Michigan, that killed six workers and injured many others was caused by a maintenance error [11, 12].
- A study of nuclear power plant operating experiences revealed that because of errors in maintenance of some motors in the rod drives, many of the motors ran in a backward direction and withdrew rods, instead of inserting them [13].

9.3 CAUSES OF HUMAN ERROR IN POWER PLANT MAINTENANCE

There are many different causes for the occurrence of human errors in power plant maintenance. On the basis of characteristics obtained from modeling the maintenance task, error causes in power plant maintenance may be classified under four major categories as shown in Figure 9.1 [1].

Design shortcomings in hardware and software include items such as deficiencies in the design of displays and controls, insufficient communication equipment, and wrong or confusing procedures. An example of human ability limitations is the limited capacity of short-term memory in the internal control mechanism.

Some important examples of disturbances of the external environment are the physical conditions such as humidity, ventilation, ambient illumination, and

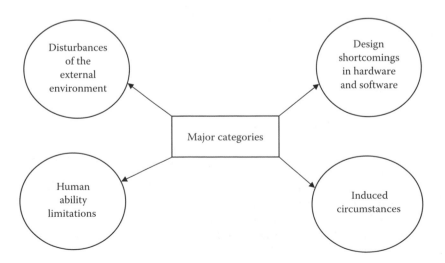

FIGURE 9.1 Major categories of error causes in power plant maintenance.

temperature. Induced circumstances include items such as momentary distractions, improper communications which may result in failures, and emergency conditions.

A study identified the following causal factors, in order of greatest to least frequency of occurrence, for critical incidents and reported events related to maintenance error in power plants [14, 15]:

- Faulty procedures
- Problems in clearing and tagging equipment for maintenance
- Shortcomings in equipment design
- Problems in moving people or equipment
- Poor training
- Poor unit and equipment identification
- Problems in facility design
- Poor work practices
- Adverse environmental factors
- Mistakes by maintenance personnel

"Faulty procedures" are the most frequently appearing causal factor in the mishaps reported. It includes items such as incorrect procedures, incompleteness, lack of specificity, and lack of adherence to a specified procedure. An example of faulty procedures is "due to poor judgment and not following prescribed guidelines properly, a ground was left on a circuit breaker. When the equipment was put back into service, the circuit breaker blew up and caused extensive property damage." In this case, the correct procedure would have required clearing the ground prior to returning the circuit breaker to service.

"Problems in clearing and tagging equipment for maintenance" are the second most frequent causal factor in reported cases where serious accidents/potentially serious accidents could be attributed to a failure/error associated with the equipment clearance process. "Shortcomings in equipment design" are the third most frequent causal factor for accidents/near-accidents revolved about equipment design-related problems. The factor includes items such as the equipment not designed with appropriate mechanical safeguards to prevent the substitution of wrong part for the proper replacement part, equipment installed incorrectly from the outset, parts placed in inaccessible locations, and poorly designed and inherently unreliable components.

"Problems in moving people or equipment" are the fourth most frequent causal factor. These problems basically stem from poor lifting capability or the inability to employ proper vehicular aids in moving heavy units of equipment. "Poor training," "poor unit and equipment identification" and "problems in facility design" are the fifth most frequent causal factors. The factor "poor training" is basically concerned with the unfamiliarity of repair workers with the job or their lack of awareness of the system characteristics and inherent dangers associated with the job at hand. "Poor unit and equipment identification" is the cause of an unexpectedly high number of accidents, and often the problem is confusion between two identical items and sometimes improper identification of potential hazards.

"Problems in facility design" can contribute to accidents. Some examples of these problems are insufficient clearances for repair workers, equipment, or transportation aids in the performance of maintenance activities, and inadequately sized facilities causing an overly dense packaging of equipment systems and preventing effective performance of repair or inspection tasks.

"Poor work practices" are the sixth most frequent causal factor. Some examples of poor work practices are not waiting for operators to complete the switching and tagging tasks essential to disable the systems requiring attention and not taking the time to erect a scaffold so that an item in midair can be accessed safely.

"Adverse environmental factors" and "mistakes by maintenance personnel" are the seventh (or the least) frequent causal factors. The "adverse environmental factors" include items such as the need to wear protective garments and devices in threatening environments that, in turn, restrict a person's movement capabilities and visual field, and the encouragement of haste by the need to minimize stay time in, say, radioactive environments. "Mistakes by maintenance personnel" are a small fraction of those errors that would be difficult to anticipate and "design-out" of power generation plants.

Additional information on all of the above causal factors is available in Ref. [14].

9.4   MAINTENANCE TASKS MOST SUSCEPTIBLE TO HUMAN ERROR IN POWER GENERATION

In the 1990s the Central Research Institute of Electric Power Industry (CRIEPI) in Japan and the Electric Power Research Institute in the United States conducted a joint study to identify critical maintenance tasks and to develop, implement, and evaluate interventions that have high potential to reduce the occurrence of human errors or increasing maintenance productivity in nuclear power plants. As the result of this study, five maintenance tasks most susceptible to the occurrence of human errors, as shown in Figure 9.2, were identified [16]. It simply means that careful attention is necessary in performing such tasks to minimize or eliminate the occurrence of human errors.

9.5   METHODS FOR PERFORMING MAINTENANCE ERROR ANALYSIS IN POWER GENERATION

Over the years, many methods or models have been developed that can be used to perform maintenance error analysis in power generation. Three such methods/ models are presented below.

9.5.1   FAULT TREE ANALYSIS

This is a widely used method in the industrial sector to perform various types of reliability-related analysis [17, 18]. The method is described in detail in Chapter 4. Its application to the performance of maintenance error analysis in the area of power generation is demonstrated through the following example:

EXAMPLE 9.1

Assume that a piece of power plant equipment can fail due to a maintenance error caused by four factors: poor work environment, carelessness, poor equipment design,

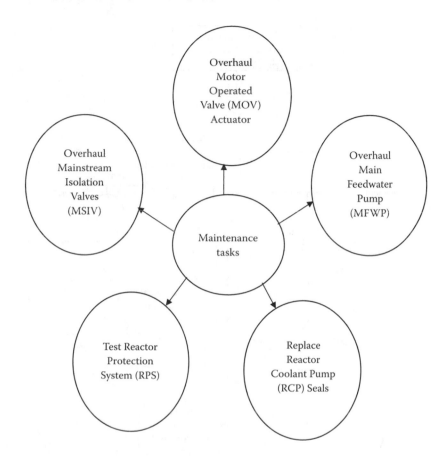

FIGURE 9.2 Maintenance tasks most susceptible to human errors.

and use of deficient maintenance manuals. Two major factors for poor work environment are inadequate lighting or distractions. Similarly, three factors for poor equipment design are oversight, misinterpretation of design specification, or no formal consideration to maintenance error occurrence in design specification. Finally, two factors for carelessness are poor training or time constraints.

Develop a fault tree for the top event "Power plant equipment failure due to a maintenance error" by using fault tree symbols given in Chapter 4.

A fault tree for the example is shown in Figure 9.3.

EXAMPLE 9.2

Assume that the probability of occurrence of events E_1, E_2, E_3, ..., E_8 shown in Figure 9.3 is 0.01. For independent events, calculate the probability of occurrence of the top event T (i.e., power plant equipment failure due to a maintenance error), and intermediate events I_1, (i.e., carelessness), I_2 (i.e., poor equipment design) and I_3 (i.e., poor work environment).

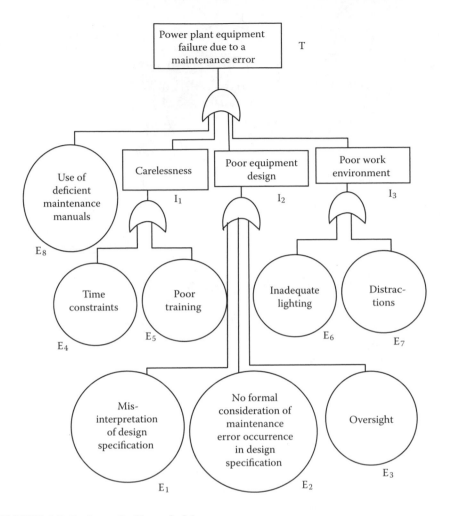

FIGURE 9.3 Fault tree for Example 9.1.

Using Chapter 4, Refs. [17, 18], and the given data, we obtain the values of I_1, I_2, I_3, and T as follows:

The probability of occurrence of event I_1 is given by

$$P(I_1) = P(E_4) + P(E_5) - P(E_4)P(E_5)$$
$$= 0.01 + 0.01 - (0.01)(0.01) \tag{9.1}$$
$$= 0.0199$$

where $P(I_1)$, $P(E_4)$, and $P(E_5)$ are the probabilities of occurrence of events I_1, E_4, and E_5, respectively.

The probability of occurrence of event I_2 is

$$
\begin{aligned}
P(I_2) &= 1 - \{1 - P(E_1)\}\{1 - P(E_2)\}\{1 - P(E_3)\} \\
&= 1 - \{1 - 0.01\}\{1 - 0.01\}\{1 - 0.01\} \\
&= 0.0297
\end{aligned}
\tag{9.2}
$$

where $P(I_2)$, $P(E_1)$, $P(E_2)$, and $P(E_3)$ are the probabilities of occurrence of events I_2, E_1, E_2, and E_3, respectively.

The probability of occurrence of event I_3 is given by

$$
\begin{aligned}
P(I_3) &= P(E_6) + P(E_7) - P(E_6)P(E_7) \\
&= 0.01 + 0.01 - (0.01)(0.01) \\
&= 0.0199
\end{aligned}
\tag{9.3}
$$

where $P(I_3)$, $P(E_6)$, and $P(E_7)$ are the probabilities of occurrence of events I_3, E_6, and E_7, respectively.

By using the above calculated and the specified values, Chapter 4, and Refs. [17, 18] we get

$$
\begin{aligned}
P(T) &= 1 - \{1 - P(E_8)\}\{1 - P(I_1)\}\{1 - P(I_2)\}\{1 - P(I_3)\} \\
&= 1 - (1 - 0.01)(1 - 0.0199)(1 - 0.0297)(1 - 0.0199) \\
&= 0.0772
\end{aligned}
\tag{9.4}
$$

Thus, the probabilities of occurrence of events T, I_1, I_2, and I_3 are 0.0772, 0.0199, 0.0297, and 0.0199, respectively.

9.5.2 MARKOV METHOD

This is a widely used method to perform reliability analysis of repairable engineering systems, and it can be used to perform maintenance error analysis in power plants. The method is described in Chapter 4. Its application to perform maintenance error analysis in the area of power generation is demonstrated through the mathematical model presented below.

This mathematical model represents a power plant system that might fail due to a maintenance error or non-maintenance error failures. The system state space diagram is shown in Figure 9.4 [19]. Numerals in boxes denote system states. The following assumptions are associated with the model:

- The system maintenance error and non-maintenance error failure rates are constant.
- The failed system is repaired and the repaired system is as good as new.
- Failed system repair rates are constant.

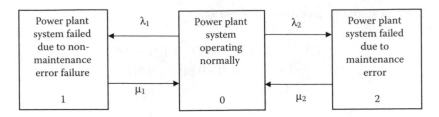

FIGURE 9.4 System state space diagram.

The following symbols are associated with the model:

i is the system state i; for $i = 0$ (power plant system operating normally),
$i = 1$ (power plant system failed due to non-maintenance error failure),
$i = 2$ (power plant system failed due to maintenance error).
$P_i(t)$ is the probability that the power plant system is in state i at time t; for
    $i = 0, 1, 2$.
λ_1 is the power plant system constant non-maintenance error failure rate.
μ_1 is the power plant system constant repair rate from state 1 to state 0.
λ_2 is the power plant system constant maintenance error rate.
μ_2 is the power plant system constant repair rate from state 2 to state 0.

By applying the Markov method described in Chapter 4, we write down the following equations for the diagram:

$$\frac{dP_0(t)}{dt} + (\lambda_1 + \lambda_2)P_0(t) = \mu_1 P_1(t) + \mu_2 P_2(t) \tag{9.5}$$

$$\frac{dP_1(t)}{dt} + \mu_1 P_1(t) = \lambda_1 P_0(t) \tag{9.6}$$

$$\frac{dP_2(t)}{dt} + \mu_2 P_2(t) = \lambda_2 P_0(t) \tag{9.7}$$

At $t = 0$, $P_0(0) = 1$, $P_1(0) = 0$, and $P_2(0) = 0$.
By solving Equations (9.5)–(9.7), we get

$$P_0(t) = \frac{\mu_1 \mu_2}{x_1 x_2} + \left[\frac{(x_1 + \mu_2)(x_2 + \mu_1)}{x_1(x_1 - x_2)}\right]e^{x_1 t} - \left[\frac{(x_2 + \mu_2)(x_2 + \mu_1)}{x_2(x_1 - x_2)}\right]e^{x_2 t} \tag{9.8}$$

where

$$x_1, x_2 = \frac{-\beta \pm \sqrt{\beta^2 - 4(\mu_2 \mu_1 + \lambda_2 \mu_1 \lambda_1 \mu_1)}}{2} \tag{9.9}$$

$$\beta = \mu_2 + \mu_1 + \lambda_1 + \lambda_2 \tag{9.10}$$

$$x_1 x_2 = \mu_1 \mu_2 + \lambda_2 \mu_1 + \lambda_1 \mu_2 \tag{9.11}$$

$$x_1 + x_2 = -(\mu_2 + \mu_1 + \lambda_2 + \lambda_1) \tag{9.12}$$

$$P_1(t) = \frac{\lambda_2 \mu_1}{x_1 x_2} + \left[\frac{\lambda_1 x_1 + \lambda_1 \mu_2}{x_1 (x_1 - x_2)} \right] e^{x_1 t} - \left[\frac{(\mu_2 + x_2)\lambda_2}{x_2 (x_1 - x_2)} \right] e^{x_2 t} \tag{9.13}$$

$$P_2(t) = \frac{\lambda_1 \mu_2}{x_1 x_2} + \left[\frac{\lambda_1 x_1 + \lambda_1 \mu_2}{x_1 (x_1 - x_2)} \right] e^{x_1 t} - \left[\frac{(\mu_2 + x_2)\lambda_1}{x_2 (x_1 - x_2)} \right] e^{x_2 t} \tag{9.14}$$

As t becomes very large, we get the following steady-state probability equations from Equations (9.8), (9.13), and (9.14), respectively:

$$P_0 = \frac{\mu_1 \mu_2}{x_1 x_2} \tag{9.15}$$

$$P_1 = \frac{\lambda_2 \mu_1}{x_1 x_2} \tag{9.16}$$

and

$$P_1 = \frac{\lambda_1 \mu_2}{x_1 x_2} \tag{9.17}$$

where P_0, P_1, and P_2 are the steady-state probabilities of the power plant system being in states 0, 1, and 2, respectively. It is to be noted that Equation (9.15) is also known as the system steady-state availability.

EXAMPLE 9.3

Assume that we have the following data values for a power plant system:

$$\lambda_1 = 0.006 \text{ failures per hour}$$

$$\lambda_2 = 0.001 \text{ errors per hour}$$

$$\mu_1 = 0.04 \text{ repairs per hour}$$

$$\mu_2 = 0.02 \text{ repairs per hour}$$

Calculate the steady-state probability of the system failing due to maintenance error. By substituting the specified data values into Equation (9.17), we get

$$P_2 = \frac{\lambda_1 \mu_2}{x_1 x_2} = \frac{\lambda_1 \mu_2}{(\mu_1 \mu_2 + \lambda_2 \mu_1 + \lambda_1 \mu_2)}$$

$$= \frac{(0.006)(0.02)}{[(0.04)(0.02) + (0.001)(0.04) + (0.006)(0.02)]}$$

$$= 0.1259$$

Thus, the steady-state probability of the power plant system failing due to maintenance error is 0.1259.

9.5.3 Maintenance Personnel Performance Simulation (MAPPS) Model

This is a computerized, stochastic, task-oriented human behavioral model developed by the Oak Ridge National Laboratory, for providing estimates of nuclear power plant (NPP) maintenance manpower performance measures [20]. Its development was sponsored by the United States Nuclear Regulatory Commission (NRC), and the primary objective for its development was the need for and lack of a human reliability-related data bank pertaining to NPP maintenance activities, for use in performing probabilistic risk assessment (PRA) studies.

The measures of performance estimated by MAPPS include the probability of successfully completing the task of interest, the task duration time, probability of an undetected error, maintenance team stress profiles during task execution, and identification of the most-and least-likely error-prone subelements. Needless to say, the MAPPS model is a powerful tool for estimating important maintenance parameters and its flexibility allows it to be useful for various applications dealing with NPP maintenance activity.

Additional information on the MAPPS model is available in Ref. [20].

9.6 STEPS FOR IMPROVING MAINTENANCE PROCEDURES IN POWER GENERATION AND USEFUL GUIDELINES FOR HUMAN ERROR REDUCTION AND PREVENTION IN POWER GENERATION MAINTENANCE

Past experiences indicate that improving maintenance procedures in power generation can help to reduce performance errors along with a corresponding increase in unit reliability. In general, the upgrade of a maintenance procedure can be accomplished by following the steps listed below [21].

- **Step 1:** This is concerned with selecting a procedure to be upgraded by considering factors such as user inputs and relative importance of the procedure.
- **Step 2:** This is concerned with reviewing the procedure with respect to items such as device nomenclature, tolerances, required test equipment, limits, step sequence, prerequisites, and precautions.
- **Step 3:** This is concerned with reviewing the procedure for agreement with the procedure development guidelines.
- **Step 4:** This is concerned with the preliminary validation of the procedure to determine its usability.
- **Step 5:** This is concerned with rewriting the procedure by taking into consideration the results of Steps 2, 3, and 4.
- **Step 6:** This is concerned with reviewing the revised procedure with respect to technical accuracy and agreement with the "Procedure Development Guide."

- **Step 7:** This is concerned with evaluating the revised procedure with respect to its usability by those responsible for performing it.
- **Step 8:** This is concerned with the approval of the upgraded procedure by appropriate supervisory and management personnel.

An upgraded maintenance procedure can substantially contribute to many areas including fewer human performance errors, identification of needed training, identification of desirable plant modifications, higher level of employee morale, and better unit reliability [21].

Additional information on improving maintenance procedures in power plants is available in Ref. [21].

Over the years, various guidelines have been proposed to reduce and prevent the occurrence of human error in power generation maintenance. Four of these guidelines are as follows [1]:

- **Revise training programs for all concerned maintenance personnel.** It basically means that training programs for maintenance personnel should be revised in accordance with the characteristics and frequency of occurrence of each extrinsic cause.
- **Ameliorate design deficiencies.** As deficiencies in design can reduce attention to the tasks and may even induce human error, this guideline calls for overcoming deficiencies in areas such as labeling, coding, plant layout, and work environment.
- **Carry out administrative policies more thoroughly.** It basically means motivating maintenance personnel appropriately to comply with prescribed quality control procedures.
- **Develop appropriate work safety checklists for maintenance personnel.** It means that maintenance personnel should be provided with work safety checklists, which can be used to determine the possibility of human error occurrence and the factors that may affect their actions prior to or after the performance of maintenance tasks.

Additional information on the above four guidelines is available in Ref. [1].

9.7 PROBLEMS

1. Write an essay on human error in power plant maintenance.
2. Discuss at least four facts and figures concerning human error in power plant maintenance.
3. What are the major causes of error in power plant maintenance?
4. Discuss power plant maintenance tasks that are most susceptible to human error.
5. Prove Equations (9.8), (9.13), and (9.14) by using Equations (9.5)–(9.7).
6. Prove that the sum of Equations (9.8), (9.13), and (9.14) is equal to unity.
7. Describe the maintenance personnel performance simulation (MAPPS) model.

8. What are the steps that can be used for improving maintenance procedures in power generation?
9. Discuss at least three useful guidelines for human error reduction and
prevention in power plant maintenance.
10. List ten causal factors in order of greatest to least frequency of occurrence,
for critical incidents and reported events directly or indirectly related to
maintenance error in power plants.

REFERENCES

1. Wu, T.M., Hwang, S.L., Maintenance Error Reduction Strategies in Nuclear Power
Plants, Using Root Cause Analysis, *Applied Ergonomics*, Vol. 20, No. 2, 1989, pp.
115–121.
2. Speaker, D.M., Voska, K.J., Luckas, W.J., Identification and Analysis of Human Errors
Underlying Electric/Electronic Component Related Events, Report No. NUREG/
CR-2987, Nuclear Power Plant Operations, United States Nuclear Regulatory
Commission, Washington, D.C., 1983.
3. WASH-1400 (NUREG-75/014), Reactor Safety Study, Report prepared by the United
States Nuclear Regulatory Commission (NRC), Washington, D.C., 1975.
4. Daniels, R.W., The Formula for Improved Plant Maintainability Must Include Human
Factors, *Proceedings of the IEEE Conference on Human Factors and Nuclear Safety*,
1985, pp. 242–244.
5. Reason, J., Human Factors in Nuclear Power Generation: A Systems Perspective,
Nuclear Europe Worldscan, Vol. 17, No. 5–6, 1997, pp. 35–36.
6. An Analysis of 1990 Significant Events, Report No. INPO 91-018, Institute of Nuclear
Power Operations (INPO), Atlanta, GA, 1991.
7. Pyy, P., An Analysis of Maintenance Failures at a Nuclear Power Plant, *Reliability
Engineering and System Safety*, Vol. 72, 2001, pp. 293–302.
8. Pyy, P., Laakso, K., Reiman, L., A Study of Human Errors Related to NPP Maintenance
Activities, *Proceedings of the IEEE 6th Annual Human Factors Meeting*, 1997,
pp. 12.23–12.28.
9. Hasegawa, T., Kameda, A., Analysis and Evaluation of Human Error Events in
Nuclear Power Plants, Presented at the Meeting of the IAEA'S CRP on "Collection
and Classification of Human Reliability Data for Use in Probabilistic Safety
Assessments," May 1998. Available from the Institute of Human Factors, Nuclear
Power Engineering Corporation, 3-17-1, Toranomon, Minato-Ku, Tokyo, Japan.
10. Maintenance Error a Factor in Blackouts, *Miami Herald*, Miami, FL, December 29,
1989, p. 4.
11. The UAW and the Rouge Explosion: A Pat on the Head, *Detroit News*, Detroit, MI,
February 6, 1999, p. 6.
12. White, J., New Revelations Expose Company-Union Complicity in Fatal Blast at US
Ford Plant. Available online at http://www.wsws.org/articles/2000/feb2000/ford-f04.
shtml.
13. Nuclear Power Plant Operating Experiences, from the IAEA/NEA Incident Reporting
System 1996–1999, Organization for Economic Co-operation and Development
(OECD), Paris, 2000.
14. Seminara, J.L., Parsons, S.O., Human Factors Review of Power Plant Maintainability,
Report No. NP-1567 (Research Project 1136), Electric Power Research Institute, Palo
Alto, CA, 1981.

15. Seminara, J.L., Parsons, S.O., Human Factors Engineering and Power Plant Maintenance, *Maintenance Management International*, Vol. 6, 1985, pp. 33–71.
16. Isoda, H., Yasutake, J.Y., Human Factors Interventions to Reduce Human Errors and Improve Productivity in Maintenance Tasks, *Proceedings of the International Conference on Design and Safety of Advanced Nuclear Power Plants*, 1992, pp. 34.4-1 to 34.4-6.
17. Dhillon, B.S., Singh, C., *Engineering Reliability: New Techniques and Applications*, John Wiley and Sons, New York, 1981.
18. Dhillon, B.S., *Human Reliability: With Human Factors*, Pergamon Press, Inc., New York, 1986.
19. Dhillon, B.S., *Engineering Maintenance: A Modern Approach*, CRC Press, Boca Raton, FL, 2002.
20. Knee, H.E., The Maintenance Personnel Performance Simulation (MAPPS) Model: A Human Reliability Analysis Tool, *Proceedings of the International Conference on Nuclear Power Plant Aging, Availability Factor and Reliability Analysis*, 1985, pp. 77–80.
21. Herrin, J.L., Heuertz, S.W., Improving Maintenance Procedures: One Utility's Perspectives, *Proceedings of the IEEE Conference on Human Factors and Power Plants*, 1988, pp. 373–377.

10 Safety in Engineering Maintenance

10.1 INTRODUCTION

Each year billions of dollars are being spent worldwide to keep engineering systems functioning effectively. The problem of safety in engineering maintenance has become an important issue because of the occurrence of various maintenance-related accidents throughout the industrial sector. For example, in 1994, in the U.S. mining sector approximately 14% of all accidents were associated with maintenance activity [1]. Since 1990, the occurrence of such accidents has been following an increasing trend [1].

The problem of safety in the area of engineering maintenance involves ensuring not only the safety of maintenance personnel but also the safety of actions taken by these individuals. Engineering maintenance activities present many unique occupation-related hazards, including performing tasks at elevated heights or with equipment/system that has significant potential for releasing mechanical or electrical energy.

All in all, engineering maintenance must strive to control or eradicate potential hazards for ensuring proper protection to individuals and material, including items such as electrical shocks, high noise levels, toxic gas sources, moving mechanical assemblies, and fire radiation sources [2, 3].

This chapter presents various important aspects of safety in engineering maintenance.

10.2 FACTS, FIGURES, AND EXAMPLES

Some of the important facts, figures, and examples that are directly or indirectly concerned with maintenance safety are presented below.

- In 1993, there were around 10,000 work-related deaths in the United States [1].
- In 1998, about 3.8 million workers suffered from disabling injuries on the job in the United States [1, 4].
- In 1994, approximately 14% of all accidents in the United States mining sector were associated with maintenance activity [1, 2].
- In 1998, the total cost of work-related injuries in the United States was estimated to be around $125 billion [1, 2, 4].
- A study of safety issues concerning onboard fatalities in jet fleets worldwide for the period 1982–1991 reported that maintenance and inspection was the second most important issue with 1481 onboard fatalities [5, 6].
- In 1985, 520 people were killed in a Japan Airlines Boeing 747 jet accident because of an incorrect repair [7, 8].

- In 1991, four workers were killed in an explosion at an oil refinery in Louisiana as three gasoline-synthesizing units were being brought back to their operating state, after going through some maintenance-related activities [9].
- In 1990, 10 people were killed on the USS *Iwo Jima* (LPH2) naval ship because of a steam leak in the fire room, after maintenance workers repaired a valve and replaced bonnet fasteners with mismatched and incorrect material [10].
- In 1979, 272 people were killed in a DC-10 aircraft accident in Chicago because of wrong procedures followed by maintenance workers [11].

10.3 CAUSES OF MAINTENANCE SAFETY PROBLEMS AND FACTORS RESPONSIBLE FOR DUBIOUS SAFETY REPUTATION IN MAINTENANCE ACTIVITY

Over the years various causes for safety problems have been identified. Some of the important ones are poor safety standards, poor work environment, poor work tools, poor training of maintenance personnel, poorly written instructions and procedures, poor management, and insufficient time to perform required maintenance tasks [2, 4].

There are many factors responsible for giving the maintenance activity a dubious safety reputation. Some of these are presented below [12].

- Performance of maintenance activities underneath or inside items such as air ducts, pressure vessels, and large rotating machines.
- Difficulty in maintaining effective communication with individuals involved in the performance of maintenance tasks.
- Sudden need for maintenance work, thus allowing a very short time for appropriate preparation.
- Disassembling previously operating equipment, thus carrying out tasks subject to the risk of releasing stored energy.
- Performance of maintenance activities in remote areas, at odd hours, and in small numbers.
- Need to carry heavy and rather bulky objects from a store/warehouse to the maintenance location, sometimes utilizing lifting and transport equipment that is way beyond the boundaries of a strict maintenance regime.
- Maintenance work performed in unfamiliar surroundings or territory imply that hazards such as missing gratings, rusted handrails, and damaged light fittings may go totally unnoticed.
- From time to time, maintenance activities may require performing tasks such as disassembling corroded parts or manhandling difficult heavy units in rather poorly lit areas and confined spaces.
- Frequent occurrence of many maintenance tasks (e.g., equipment failures), thus lesser opportunity for discerning safety-related problems and for initiating appropriate remedial actions.

10.4 FACTORS INFLUENCING SAFETY BEHAVIOR AND SAFETY CULTURE IN MAINTENANCE PERSONNEL

There are many factors that influence safety behavior and safety culture in mainte-nance personnel. For example, some of the factors that influence safety behavior and safety culture in railway maintenance workers are as follows [13]:

- Poor and underutilized real-time risk assessment skills
- Communication on the job (poor quality and excessive)
- Individual perception of what "safe" is
- Management personnel's communication methods
- Feedback messages from management personnel
- Physical conditions
- Supervisory personnel's visibility and accessibility
- Volume of paper work
- Reporting methods
- Equipment (condition, appropriateness, and availability)
- Competence capability and certification
- Fatigue, concentration, and ability to function
- Peer pressure
- Practical alternatives to rules
- Inconsistent teams
- Contradictory rules
- Perceived objective of the rule book
- Rule dissemination
- Training methods and training needs analysis
- Safety role model behavior
- Perceived purpose of paper work
- Pre-job information dissemination
- Rule book usability and availability
- Social pressure of home life

10.5 GOOD SAFETY-RELATED PRACTICES DURING MAINTENANCE WORK AND MAINTENANCE-RELATED SAFETY MEASURES CONCERNING MACHINERY

It is very important to follow good practices before, during, and after maintenance operations because of the existence of various types of hazards. Failure to follow good practices during any phase of maintenance can lead to potentially hazardous conditions. Four good safety-related practices to be followed during maintenance work are as follows [14].

- **Prepare for Maintenance during the Design Phase**
  It basically means that preparation for maintenance actually starts during the design of the facility by ensuring that appropriate indicators are in place for allowing effective troubleshooting and diagnostic work. Furthermore,

the equipment is designed so that normal safety-related measures can easily be taken before the maintenance activity. More specifically, equipment is designed so that all appropriate safeguards are in place for allowing it to be drained, purged, isolated, and analyzed effectively.

- **Prepare All Staff Members for Maintenance Operations**
Usually maintenance activity involves opening equipment that contains hazardous material during its normal operations. Thus, it is important to take necessary precautions prior to working on such equipment to ensure that it is completely free from residual material and is at a safe temperature and pressure. Often equipment is prepared for maintenance by people other than those actually performing maintenance on the equipment.

  In this scenario, it is essential to prepare all staff members (i.e., who prepare the equipment for maintenance and the others who perform maintenance) for maintenance operations.

- **Highlight All Potential Hazards and Plan Effectively Well in Advance**
There is no substitute for proper job planning as effective equipment isolation prior to the maintenance activity starts with thorough preplanning. Also, good practice guidelines clearly state that all potential hazards are most effectively recognized during the planning process, rather than during the job execution in a stressful environment.

  In summary, ensure that the equipment under consideration is properly freed from all types of potential hazards and that all safety precautions can be satisfied effectively. In situations when procedures cannot be followed effectively and/or safety precautions cannot be fully satisfied, do not proceed any further until a proper hazard evaluation can be carried out and a safe course of measures determined.

- **Plan Now for the Future**
This is concerned with analyzing the potential effects on the maintenance activity when changes are made to the existing process. Along with the determination of how operations will be affected, process management must carefully evaluate questions such as: Will there be need for more frequent or less frequent maintenance? Will maintenance personnel be at greater risk because of this change? and How will this change affect all the future maintenance-related activities?

  Over the years safety specialists have done much to point out various safety measures to be observed in working around machinery, particularly with respect to the maintenance activity. Past experiences indicate that all of these and the application of careful planning have considerably reduced the occurrence of accidents and damage to machinery. The following maintenance-related safety measures have proven to be very useful [15]:

  - All types of machines properly equipped with appropriate safety valves, alarms for indicating abnormal operating conditions, and over-speed cutouts
  - Appropriate guards around exposed moving parts of machining equipment

- Platforms, ladders, and stairways with appropriate protective features
- Safety shoes, hats, gloves, and clothing
- Items such as portable electric drills, grinders, and electric motors should have proper ground wire attached to prevent maintenance workers and others coming in contact with defective wiring on machining equipment
- Equipment designed for work intended should an appropriate level of safe margin for insuring safe operation under extreme environments
- Safe tools for clipping and grinding and appropriate goggles for eye protection
- All types of electrical equipment installed according to currently approved code

10.6   MAINTENANCE SAFETY-RELATED QUESTIONS FOR ENGINEERING EQUIPMENT MANUFACTURERS

Engineering equipment manufacturers can play a key role in improving maintenance safety during equipment field use by effectively addressing common problems that might be encountered during the maintenance activity. Questions such as the ones presented below can be quite useful to equipment manufacturers in determining whether the common problems that might be encountered during the equipment maintenance activity have been addressed properly [16].

- Are all the test points located at easy to find and reach locations?
- Are the components requiring frequent maintenance easily accessible all the time?
- Are effectively written instructions available for repair and maintenance activities?
- Can the disassembled piece of equipment for repair be reassembled incorrectly so that it becomes hazardous to all potential users?
- Were human factors principles properly applied to reduce maintenance problems?
- Is the repair process hazardous to all involved repair workers?
- Do the repair instructions contain effective warnings to wear appropriate gear because of pending hazards?
- Are warnings properly placed on parts that can shock maintenance personnel?
- Is the need for special tools for repairing safety-critical parts reduced to a minimum level?
- Is there a proper system to remove hazardous fluid from the equipment/ system to be repaired?
- Does the equipment contain proper safety interlocks that must be bypassed for performing essential repairs/adjustments?
- Is the equipment/system designed in such a way that after a failure, it would automatically stop operating and would cause absolutely no damage?
- Does the equipment contain an appropriate built-in system to indicate that safety-critical parts need maintenance?

- Do the instructions include warnings for alerting maintenance personnel of any danger?
- Is there an appropriate mechanism installed for indicating when the backup units of safety-critical systems fail?
- Was proper attention given to reducing voltages to levels at test points so that hazards to maintenance workers are reduced?

10.7 GUIDELINES FOR ENGINEERING EQUIPMENT DESIGNERS TO IMPROVE SAFETY IN MAINTENANCE

Over the years, professionals working in the area of maintenance have developed various guidelines for engineering equipment designers, considered useful to improve safety in maintenance. Some of these guidelines are presented below [16].

- Pay close attention to typical human behaviors and eliminate or reduce the need for special tools.
- Install appropriate interlocks for blocking access to hazardous locations and provide effective guards against moving parts.
- Develop designs/procedures in such a way that the maintenance error occurrence probability is reduced to a minimum.
- Design for easy accessibility so that parts requiring maintenance are easy and safe to check, replace, service, or remove.
- Incorporate effective fail-safe designs to prevent damage or injury in the event of a failure.
- Eliminate or reduce the need to perform adjustments/maintenance close to hazardous operating parts.
- Incorporate appropriate devices/measures for early detection or prediction of all types of potential failures so that necessary maintenance can be carried out prior to actual failure with a reduced risk of hazards.
- Develop the design in such a way that the probability of maintenance workers being injured by escaping high-pressure gas, electric shock, and so on, is reduced to a minimum.

10.8 MATHEMATICAL MODELS

Over the years, a large number of mathematical models have been developed to perform various types of reliability and availability analysis of engineering systems [17]. Some of these models can also be used to perform maintenance safety-related analysis of engineering systems. One such model is presented below.

This mathematical model represents an engineering system with three states: operating normally, working unsafely (due to maintenance or other problems), and failed. The system is repaired from failed and unsafe working states. The system state space diagram is shown in Figure 10.1. The numerals in boxes and circle denote system states. The Markov method described in Chapter 4 is used to develop equations for system state probabilities and mean time to failure.

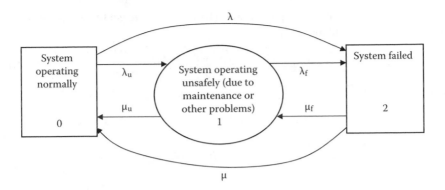

FIGURE 10.1 System state space diagram.

The following assumptions are associated with the model:

- All occurrences are independent of each other.
- System failure and repair rates are constant.
- The repaired system is as good as new.

The following symbols are associated with the diagram:

i is the ith state of the system: $i = 0$ (system operating normally), $i = 1$ (system operating unsafely due to maintenance or other problems), $i = 2$ (system failed).

t is time.

$P_i(t)$ is the probability that the system is in state i at time t; for $i = 0, 1, 2$.

λ is the system constant failure rate.

λ_u is the system constant unsafe degradation rate due to maintenance or other problems.

λ_f is the system constant failure rate from its unsafe operating state 1.

μ is the system constant repair rate from state 2 to state 0.

μ_u is the system constant repair rate from state 1 to state 0.

μ_f is the system constant repair rate from state 2 to state 1.

Using the Markov method, we write down the following equations for the diagram [2, 17]:

$$\frac{dP_0(t)}{dt} + (\lambda_u + \lambda)P_0(t) = \mu_u P_1(t) + \mu P_2(t) \tag{10.1}$$

$$\frac{dP_1(t)}{dt} + (\mu_u + \lambda_f)P_1(t) = \mu_u P_2(t) + \lambda_u P_0(t) \tag{10.2}$$

$$\frac{dP_2(t)}{dt} + (\mu_u + \mu_f)P_2(t) = \lambda_f P_1(t) + \lambda P_0(t) \tag{10.3}$$

At time $t = 0$, $P_0(0) = 1$, $P_1(0) = 0$, and $P_2(0) = 0$.

For a very large t, by solving Equations (10.1)–(10.3), we get the following steady-state probability equations [17]:

$$P_0 = \frac{(\mu + \mu_f)(\mu_u + \lambda_f) - \lambda_f \mu_f}{X} \tag{10.4}$$

where

$$X = (\mu + \mu_f)(\mu_u + \lambda_u + \lambda_f) + \lambda(\mu_u + \lambda_f) + \lambda\,\mu_f + \lambda_u\,\lambda_f - \lambda_f\,\mu_f$$

$$P_1 = \frac{\lambda_u(\mu + \mu_f) + \lambda\mu_f}{X} \tag{10.5}$$

$$P_2 = \frac{\lambda\lambda_f + \lambda(\mu_u + \lambda_f)}{X} \tag{10.6}$$

where P_0, P_1, and P_2 are the steady-state probabilities of the system being in states 0, 1, and 2, respectively.

Thus, the steady-state probability of the system operating unsafely due to maintenance or other problems is given by Equation (10.5).

By setting $\mu = \mu_f = 0$ in Equations (10.1)–(10.3) and solving the resulting equations, we get the following equation for the system reliability:

$$\begin{aligned} R_S(t) &= P_0(t) + P_1(t) \\ &= (X_1 + Y_1)e^{x_1 t} + (X_2 + Y_2)e^{x_2 t} \end{aligned} \tag{10.7}$$

where $R_S(t)$ is the system reliability at time t.

$$x_1 = \frac{-L_1 + \sqrt{L_1^2 - 4L_2}}{2} \tag{10.8}$$

$$x_2 = \frac{-L_1 - \sqrt{L_1^2 - 4L_2}}{2} \tag{10.9}$$

$$L_1 = \mu_u + \lambda + \lambda_u + \lambda_f \tag{10.10}$$

$$L_2 = \lambda\,\mu_u + \lambda\,\lambda_f + \lambda_u\,\lambda_f \tag{10.11}$$

$$X_1 = \frac{x_1 + \mu_u + \lambda_f}{(x_1 - x_2)} \tag{10.12}$$

$$X_2 = \frac{x_2 + \mu_u + \lambda_f}{(x_2 - x_1)} \tag{10.13}$$

$$Y_1 = \frac{\lambda_u}{(x_1 - x_2)} \tag{10.14}$$

$$Y_2 = \frac{\lambda_u}{(x_2 - x_1)} \tag{10.15}$$

By integrating Equation (10.7) over the time interval $[0, \infty]$, we obtain the following equation for the system mean time to failure with repair [2, 17]:

$$
\begin{aligned}
MTTF_{Sr} &= \int_0^\infty R_S(t)\,dt \\
&= \left[\frac{(X_1 + Y_1)}{x_1} + \frac{(X_2 + Y_2)}{x_2} \right]
\end{aligned}
\tag{10.16}
$$

where $MTTF_{Sr}$ is the system mean time to failure with repair.

EXAMPLE 10.1

Assume that a repairable engineering system can be either operating normally, operating unsafely due to maintenance or other problems, or failed. Its constant failure/degradation rates from normal operating state to failed state, normal working state to unsafe operating state, and unsafe operating state to failed state are 0.004 failures per hour, 0.002 failures per hour, and 0.001 failures per hour, respectively.

Similarly, the system constant repair rates from the failed state to normal operating state, unsafe operating state to normal operating state, and failed state to unsafe working state are 0.008 repairs per hour, 0.005 repairs per hour, and 0.002 repairs per hour, respectively.

Calculate the steady-state probability of the system being in unsafe operating state due to maintenance or other problems.

By substituting the given data values into Equation (10.5), we get

$$
\begin{aligned}
P_1 &= \frac{(0.002)(0.008 + 0.002) + (0.004)(0.002)}{X} \\
&= 0.25
\end{aligned}
$$

where

$$
X = (0.008 + 0.002)(0.005 + 0.002 + 0.001) + 0.004(0.005 + 0.001)
$$

$$
+ (0.004)(0.002) + (0.002)(0.001) - (0.001)(0.002)
$$

Thus, the steady-state probability of the system being in unsafe operating state due to maintenance or other problems is 0.25.

10.9 PROBLEMS

1. Write an essay on safety in engineering maintenance.
2. List at least six facts, figures, and examples directly or indirectly concerned with safety in engineering maintenance.
3. What are the important causes of maintenance safety problems?

4. What are the factors responsible for dubious safety reputation in maintenance activity?
5. Discuss the factors influencing safety behavior and safety culture in maintenance personnel.
6. Discuss at least four good safety-related practices during maintenance work.
7. Discuss maintenance-related safety measures concerning machinery.
8. Write down at least ten maintenance safety-related questions for engineering equipment manufacturers.
9. Prove Equations (10.4)–(10.6) by using Equations (10.1)–(10.3).
10. Assume that an engineering system can be either operating normally,
operating unsafely due to maintenance or other problems, or failed. Its
constant failure/degradation rates from normal operating state to failed
state, normal working state to unsafe operating sate, and unsafe operating
state to failed state are 0.003 failures per hour, 0.001 failures per hour, and
0.002 failures per hour, respectively. Similarly, the system constant repair
rates from the failed state to normal operating state, unsafe operating state
to normal operating state, and failed state to unsafe working state are
0.007 repairs per hour, 0.006 repairs per hour, and 0.001 repairs per hour,
respectively. Calculate the steady state probability of the system being in
unsafe operating state due to maintenance or other problems.

REFERENCES

1. *Accident Facts*, National Safety Council, Chicago, IL, 1999.
2. Dhillon, B.S., *Engineering Safety: Fundamentals, Techniques, and Applications*,
World Scientific Publishing, River Edge, NJ, 2003.
3. AMCP 706-132, Maintenance Engineering Techniques, U.S. Army Material Command,
Department of the Army, Washington, D.C., 1975.
4. Dhillon, B.S., *Engineering Maintenance: A Modern Approach*, CRC Press, Boca
Raton, FL, 2002.
5. Human Factors in Airline Maintenance: A Study of Incident Reports, Bureau of Air
Safety Investigation, Department of Transport and Regional Development, Canberra,
Australia, 1997.
6. Russell, P.D., Management Strategies for Accident Prevention, *Air Asia*, Vol. 6, 1994,
pp. 31–41.
7. Gero, D., *Aviation Disasters*, Patrick Stephens, Sparkford, UK, 1993.
8. ATSB Survey of Licensed Aircraft Maintenance Engineers in Australia, Report No.
ISBN 0642274738, Australian Transport Safety Bureau (ATSB), Department of
Transport and Regional Services, Canberra, Australia, 2001.
9. Goetsch, D.L., *Occupational Safety and Health*, Prentice-Hall, Englewood Cliffs, NJ,
1996.
10. *Joint Fleet Maintenance Manual, Vol. 5, Quality Assurance, Submarine Maintenance
Engineering*, United States Navy, Portsmouth, NH, 1991.
11. Christensen, J.M., Howard, J.M., Field Experience in Maintenance, in *Human
Detection and Diagnosis of System Failures*, edited by J. Rasmussen and W.B. Rouse,
Plenum Press, New York, 1981, pp. 111–133.
12. Stoneham, D., *The Maintenance Management and Technology Handbook*, Elsevier
Science, Oxford, UK, 1998.

13. Farrington-Darby, T., Pickup, L., Wilson, J.R., Safety Culture in Railway Maintenance, *Safety Science*, Vol. 43, 2005, pp. 39–60.

14. Wallace, S.J., Merritt, C.W., Know When to Say "When": A Review of Safety Incidents Involving Maintenance Issues, *Process Safety Progress*, Vol. 22, No. 4, 2003, pp. 212–219.

15. Pender, W.R., Safety in Maintenance, *Southern Power and Industry*, Vol. 62, No. 12, 1944, pp. 98, 99, and 110.

16. Hammer, W., *Product Safety Management and Engineering*, Prentice-Hall, Englewood Cliffs, NJ, 1980.

17. Dhillon, B.S., *Design Reliability: Fundamentals and Applications*, CRC Press, Boca Raton, FL, 1999.

11 Mathematical Models for Performing Human Reliability and Error Analysis in Engineering Maintenance

11.1 INTRODUCTION

Mathematical modeling is a widely used approach to perform various types of analysis in engineering systems. In this case, the components of a system are denoted by idealized elements assumed to have representative characteristics of real-life components and whose behavior can be described by equations. However, the degree of realism of mathematical models depends on the type of assumptions imposed on them.

Over the years, a large number of mathematical models have been developed to study human reliability and error in engineering systems. Most of these models were developed using stochastic processes including the Markov approach [1, 2]. Although the usefulness of such models can vary from one situation to another, some of the human reliability and error models are being used quite successfully to represent various types of real-life environments in the industrial sector [3]. Thus, some of these models can also be used to tackle human reliability and error problems in the area of engineering maintenance.

This chapter presents the mathematical models considered quite useful to perform various types of human reliability and error-related analysis in engineering maintenance.

11.2 MODELS FOR PREDICTING MAINTENANCE PERSONNEL RELIABILITY IN NORMAL AND FLUCTUATING ENVIRONMENTS

Maintenance personnel perform various types of time-continuous tasks including monitoring, tracking, and operating. Environments under which such tasks are performed can be either normal or fluctuating. In performing such tasks, maintenance personnel can make various types of errors including critical and noncritical errors. Therefore, this section presents three mathematical models to predict maintenance worker performance reliability and to perform maintenance error-related analysis under the above-described conditions.

11.2.1   MODEL I

This model is concerned with predicting the maintenance worker performance reliability under normal conditions—more specifically, the probability of performing a time-continuous task correctly by a maintenance worker. An expression to predict the maintenance worker performance reliability is developed below [1, 2, 4, 5].

The probability of human error in a maintenance task in the finite time interval Δt with event D given is expressed by

$$P(C/D) = z(t)\Delta t \tag{11.1}$$

where C is an event that human error will occur in time interval $[t, t + \Delta t]$, D is an errorless performance event of duration t, and $z(t)$ is the human error rate at time t.

The joint probability of the errorless performance is given by

$$P(\bar{C}/D) = P(D) - P(C/D)P(D) \tag{11.2}$$

where $P(D)$ is the occurrence probability of event D and $\bar{C}$ is the event that human error will not occur in time interval $[t, t + \Delta t]$.

Equation (11.2) denotes an errorless performance probability over time intervals $[0, t]$ and $[t, t + \Delta t]$ and is rewritten as

$$R_h(t) - R_h(t)P(C/D) = R_h(t + \Delta t) \tag{11.3}$$

where $R_h(t)$ is the maintenance worker reliability at time t.

By substituting Equation (11.1) into Equation (11.3), we get

$$\lim_{\Delta t \to 0} \frac{R_h(t + \Delta t) - R_h(t)}{\Delta t} = -R_h(t)z(t) \tag{11.4}$$

In the limiting case Equation (11.4) becomes

$$\lim_{\Delta t \to 0} \frac{R_h(t + \Delta t) - R_h(t)}{\Delta t} = \frac{dR_h(t)}{dt} = -R_h(t)z(t) \tag{11.5}$$

At time $t = 0$, $R_h(0) = 1$.

By rearranging Equation (11.5), we get

$$\frac{1}{R_h(t)} \cdot dR_h(t) = -z(t)dt \tag{11.6}$$

Integrating both sides of Equation (11.6) over the time interval $[0, t]$, we get

$$\int_{1}^{R_h(t)} \frac{1}{R_h(t)} \cdot dR_h(t) = -\int_{0}^{t} z(t)dt \tag{11.7}$$

After evaluating Equation (11.7) we obtain

$$R_h(t) = -e^{-\int_0^t z(t)dt} \tag{11.8}$$

Equation (11.8) is the general expression to compute maintenance worker performance reliability for any time to human error statistical distribution (e.g., Weibull, normal, and exponential).

By integrating Equation (11.8) over the time interval $[0, \infty]$, we get the following general equation for the mean time to human error [1]:

$$MTTHE = \int_{0}^{\infty} \left[e^{-\int_0^t z(t)dt} \right] dt \tag{11.9}$$

where *MTTHE* is the mean time to human error of a maintenance worker.

EXAMPLE 11.1

Assume that a maintenance worker is performing a certain task and his or her error rate is 0.001 errors/hour (i.e., times to human error are exponentially distributed). Calculate the maintenance worker's reliability during a 6-hour work period.

Thus, we have [1]

$$z(t) = 0.001 \text{ errors/hour}$$

By substituting the above value and the given value for time t into Equation (11.8), we get

$$R_h = (6) = e^{-\int_0^6 (0.001)dt}$$
$$= e^{-(0.001)(6)}$$
$$= 0.9940$$

Thus, the maintenance worker's reliability during the 6-hour work period is 0.9940.

11.2.2 MODEL II

This model represents a maintenance worker performing time-continuous tasks under fluctuating environment (i.e., normal and stressful) [1, 6]. One example of such an environment is weather changing from normal to stormy and vice versa.

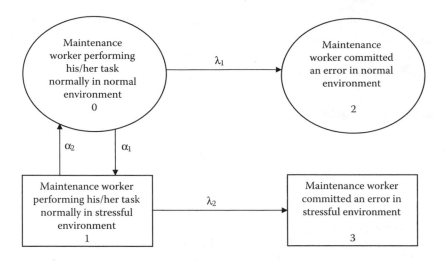

FIGURE 11.1 State space diagram for model II.

As the rate of a maintenance worker's errors from a normal work environment to a stressful environment can vary quite significantly, the model considers two separate maintenance worker error rates (i.e., one for normal environment and the other for stressful environment).

Thus, the model can be used to determine the maintenance worker's reliability and mean time to human error under the fluctuating environment. The model state space diagram is shown in Figure 11.1. The numerals in circles and boxes denote the maintenance worker's states.

The following assumptions are associated with the model:

- Maintenance worker error rates are constant.
- All maintenance worker errors occur independently.
- Environment change rates (i.e., from normal to stressful and vice versa) are constant.

The following symbols are associated with the diagram:

i is the ith state of the maintenance worker; $i = 0$ (maintenance worker performing his or her task normally in a normal environment), $i = 1$ (maintenance worker performing his or her task normally in a stressful environment), $i = 2$ (maintenance worker committed an error in a normal environment), $i = 3$ (maintenance worker committed an error in a stressful environment).

$P_i(t)$ is the probability of the maintenance worker being in state i at time t, for $i = 0, 1, 2, 3$.

λ_1 is the constant error rate of the maintenance worker performing his or her task in a normal environment.

λ_2 is the constant error rate of the maintenance worker performing his or her task in a stressful environment.

α_1 is the constant transition rate from normal environment to stressful environment.

α_2 is the constant transition rate from stressful environment to normal environment.

Using the Markov approach described in Chapter 4, we write down the following set of equations for the diagram shown in Figure 11.1 [6]:

$$\frac{dP_0(t)}{dt} + (\lambda_1 + \alpha_1) P_0(t) = \alpha_2 P_1(t) \tag{11.10}$$

$$\frac{dP_1(t)}{dt} + (\lambda_2 + \alpha_2) P_1(t) = \alpha_1 P_0(t) \tag{11.11}$$

$$\frac{dP_2(t)}{dt} = \lambda_1 P_0(t) \tag{11.12}$$

$$\frac{dP_3(t)}{dt} = \lambda_2 P_1(t) \tag{11.13}$$

At time $t = 0$, $P_0(0) = 1$, $P_1(0) = P_2(0) = P_3(0) = 0$.

By solving Equations (11.10)–(11.13), we get the following state probability equations:

$$P_0(t) = \frac{1}{(y_2 - y_1)} [(y_2 + \lambda_2 + \alpha_2) e^{y_2 t} - (y_1 + \lambda_2 + \alpha_2) e^{y_1 t}] \tag{11.14}$$

where

$$y_1 = \left[-a_1 + \left(a_1^2 - 4a_2 \right)^{1/2} \right]/2 \tag{11.15}$$

$$y_2 = \left[-a_1 - \left(a_1^2 - 4a_2 \right)^{1/2} \right]/2 \tag{11.16}$$

$$a_1 = \lambda_1 + \lambda_2 + \alpha_2 + \alpha_1 \tag{11.17}$$

$$a_2 = \lambda_1 (\lambda_2 + \alpha_2) + \alpha_1 \lambda_2 \tag{11.18}$$

$$P_1(t) = a_4 + a_5 e^{y_2 t} - \alpha_6 e^{y_1 t} \tag{11.19}$$

where

$$a_3 = \frac{1}{y_2 - y_1} \tag{11.20}$$

$$a_4 = \lambda_1 (\lambda_2 + \alpha_2)/y_1 \, y_2 \tag{11.21}$$

$$a_5 = a_3 (\lambda_1 + a_4 y_1) \tag{11.22}$$

$$a_6 = a_3 (\lambda_1 + a_4 y_2) \tag{11.23}$$

$$P_1(t) = \alpha_1 a_3 (e^{y_2 t} - e^{y_1 t}) \tag{11.24}$$

$$P_3(t) = a_7 \left[(1 + a_3)(y_1 e^{y_2 t} - y_2 e^{y_1 t}) \right] \tag{11.25}$$

where

$$a_7 = \lambda_2 \, \alpha_1 / y_1 \, y_2 \tag{11.26}$$

The maintenance worker's reliability is expressed by

$$R_{mw}(t) = P_0(t) + P_1(t) \tag{11.27}$$

where $R_{mw}(t)$ is the maintenance worker's reliability of performing tasks in fluctuating environments.

The maintenance worker's mean time to human error is given by

$$MTTHE_{mw} = \int_0^\infty R_{mw}(t) dt$$
$$= \frac{\lambda_2 + \alpha_1 + \alpha_2}{\lambda_1 (\lambda_2 + \alpha_2) + \alpha_1 \lambda_2} \tag{11.28}$$

where $MTTHE_{mw}$ is the mean time to human error of the maintenance worker performing his or her task in a fluctuating environment.

EXAMPLE 11.2

Assume that a maintenance worker's constant error rates in normal and stressful environments are 0.0001 errors/hour and 0.0005 errors/hour, respectively. The values of the transition rates from normal to stressful environment and vice versa are 0.002 times per hour and 0.003 times per hour, respectively. Calculate the mean time to human error of the maintenance worker.

Substituting the given data values into Equation (11.28) yields

$$MTTHE_{mw} = \frac{0.0005 + 0.002 + 0.003}{0.0001(0.0005 + 0.003) + (0.002)(0.0005)}$$
$$= 4074.1 \text{ hours}$$

Thus, the mean time to human error of the maintenance worker is 4074.1 hours.

11.2.3 MODEL III

This model represents a maintenance worker performing a time-continuous task subjected to critical and noncritical errors. The model can be used to calculate the maintenance worker reliability at time t, the maintenance worker mean time to human error, the probability of the maintenance worker committing a critical error at time t, and the probability of the maintenance worker committing a noncritical error at time t.

The model state space diagram is shown in Figure 11.2. The numerals in the boxes denote the maintenance worker's states.

The model is subjected to the following assumptions:

- All maintenance worker errors occur independently.
- Maintenance worker critical and noncritical error rates are constant.

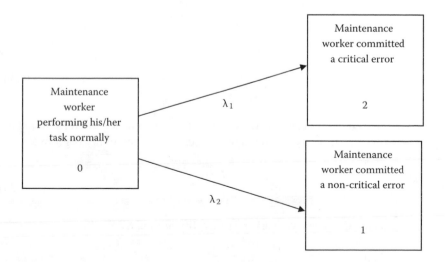

FIGURE 11.2 State space diagram for model III.

The following symbols are associated with the diagram:

i is the ith state of the maintenance worker; $i = 0$ (maintenance worker performing his or her task normally), $i = 1$ (maintenance worker committed a
noncritical error), $i = 2$ (maintenance worker committed a critical error).

$P_i(t)$ is the probability of the maintenance worker being in state i at time t, for
$i = 0, 1, 2$.

λ_1 is the constant critical human error rate of the maintenance worker.

λ_2 is the constant noncritical human error rate of the maintenance worker.

Using the Markov method, we write down the following equations for the
diagram [1, 7]:

$$\frac{dP_0(t)}{dt} + (\lambda_2 + \lambda_1) P_0(t) = 0 \tag{11.29}$$

$$\frac{dP_1(t)}{dt} - \lambda_2 P_0(t) = 0 \tag{11.30}$$

$$\frac{dP_2(t)}{dt} - \lambda_1 P_0(t) = 0 \tag{11.31}$$

At time $t = 0$, $P_0(0) = 1$, $P_1(0) = 0$, and $P_2(0) = 0$.

Solving Equations (11.29)–(11.31), we obtain the following equations:

$$P_0(t) = e^{-(\lambda_2 + \lambda_1)t} \tag{11.32}$$

$$P_1(t) = \frac{\lambda_2}{\lambda_1 + \lambda_2}[1 - e^{-(\lambda_2 + \lambda_1)t}] \tag{11.33}$$

$$P_2(t) = \frac{\lambda_1}{\lambda_1 + \lambda_2}[1 - e^{-(\lambda_2 + \lambda_1)t}] \tag{11.34}$$

The above three equations can be used to obtain the maintenance worker's probabilities of being in state 0, 1, and 2. The maintenance worker reliability is given by

$$\begin{aligned} R_m(t) &= P_0(t) \\ &= e^{-(\lambda_2 + \lambda_1)t} \end{aligned} \tag{11.35}$$

where $R_m(t)$ is the maintenance worker's reliability at time t.

The mean time to human error of the maintenance worker is given by [1, 7].

$$MTTHE_m = \int_0^\infty R_m(t)dt$$

$$= \int_0^\infty e^{-(\lambda_2 + \lambda_1)t}dt \qquad (11.36)$$

$$= \frac{1}{\lambda_2 + \lambda_1}$$

where $MTTHE_m$ is the mean time to human error of the maintenance worker.

EXAMPLE 11.3

Assume that a maintenance worker is performing a time-continuous task and his or her constant critical and noncritical error rates are 0.0001 errors/hour and 0.0006 errors/hour, respectively. Calculate the maintenance worker's reliability for a 6-hour mission and mean time to human error.

By substituting the given data values into Equations (11.35) and (11.36), we obtain

$$R_m(6) = e^{-(0.0006+0.0001)(6)}$$
$$= 0.9958$$

and

$$MTTHE_m = \frac{1}{0.0006 + 0.0001}$$
$$= 1428.6 \text{ hours}$$

Thus, the maintenance worker's reliability and mean time to human error are 0.9958 and 1428.6 hours, respectively.

11.3 MODELS FOR PERFORMING SINGLE SYSTEMS MAINTENANCE ERROR ANALYSIS

Past experiences indicate that systems can fail or degrade due to maintenance errors. Over the years, various mathematical models have been developed to perform reliability and availability analysis of such systems [1, 3, 7]. Two of these models are presented below.

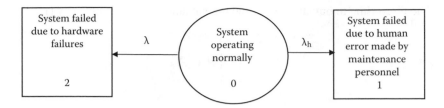

FIGURE 11.3 State space diagram for model I.

11.3.1 MODEL I

This model represents a system that can fail either due to human errors made by maintenance personnel or due to hardware failures. The model state space diagram is shown in Figure 11.3 where the numerals in the circle and boxes denote system states. It is to be noted that mathematically this model is the same as model III in Section 11.2 above, but its application is different.

The following two assumptions are associated with the model:

- Hardware failures and human errors occur independently.
- Both hardware failure and human error rates are constant.

The following symbols are associated with the diagram:

λ is the constant hardware failure rate of the system.
λ_h is the constant human error rate of the maintenance personnel.
j is the jth state of the system; $j = 0$ (system operating normally), $j = 1$ (system failed due to human error made by maintenance personnel), $j = 2$ (system failed due to hardware failures).
$P_j(t)$ is the probability of the system being in state j at time t, for $j = 0, 1, 2$.

By using the Markov method, we write down the following three equations for the diagram [1, 7]:

$$\frac{dP_0(t)}{dt} + (\lambda_h + \lambda)P_0(t) = 0 \tag{11.37}$$

$$\frac{dP_1(t)}{dt} - \lambda_h P_0(t) = 0 \tag{11.38}$$

$$\frac{dP_2(t)}{dt} - \lambda P_0(t) = 0 \tag{11.39}$$

At time $t = 0$, $P_0(0) = 1$, $P_1(0) = 0$, and $P_2(0) = 0$.

By solving Equations (11.37)–(11.39), we get

$$P_0(t) = e^{-(\lambda_h + \lambda)t} \tag{11.40}$$

$$P_1(t) = \frac{\lambda_h}{\lambda_h + \lambda}[1 - e^{-(\lambda_h + \lambda)t}] \tag{11.41}$$

$$P_2(t) = \frac{\lambda}{\lambda_h + \lambda}[1 - e^{-(\lambda_h + \lambda)t}] \tag{11.42}$$

The system reliability is given by

$$R_S(t) = P_0(t) \\ = e^{-(\lambda_h + \lambda)t} \tag{11.43}$$

where $R_S(t)$ is the system reliability at time t.

The system mean time to failure is expressed by

$$MTTF_S = \int_0^\infty R_S(t)dt \\ = \int_0^\infty e^{-(\lambda_h + \lambda)t}dt \tag{11.44} \\ = \frac{1}{\lambda_h + \lambda}$$

where $MTTF_S$ is the system mean time to failure.

EXAMPLE 11.4

Assume that a system can fail either due to human error made by maintenance personnel or due to hardware failures. The system constant human error and hardware failure rates are 0.0001 errors/hour and 0.0009 failures/hour, respectively.

Calculate the probability that the system will fail due to a human error made by maintenance personnel during a 12-hour mission. By substituting the specified data values into Equation (11.41), we obtain

$$P_1(12) = \frac{0.0001}{(0.0001 + 0.0009)}[1 - e^{-(0.0001 + 0.0009)(12)}] \\ = 0.0012$$

Thus, the probability that the system will fail due to a human error made by maintenance personnel is 0.0012.

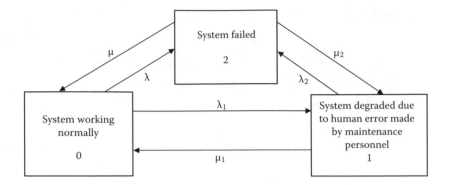

FIGURE 11.4 State space diagram for model II.

11.3.2 MODEL II

This model represents a system that can only fail due to hardware failures, but human errors made by maintenance personnel can degrade its performance.

The system is repaired from failed and degraded states. The system state space diagram is shown in Figure 11.4. The numerals in boxes denote system states.

The following assumptions are associated with the model:

- The occurrence of human error by maintenance personnel can only result in system degradation, but not failure.
- Human error and hardware failure rates are constant.
- The totally or partially failed system is repaired and preventive maintenance is performed on a regular basis.
- The degraded system can only fail due to hardware failures.
- All system repair rates are constant.
- The repaired system is as good as new.

The following symbols are associated with the diagram:

λ_1 is the constant human error rate due to maintenance personnel.
λ_2 is the system constant failure rate from its degraded state.
λ is the system constant failure rate.
μ is the system constant repair rate.
μ_1 is the constant repair rate from the system degraded state to normal working state.
μ_2 is the constant repair rate from the system failed state to degraded or partially working state.
j is the jth state of the system; $j = 0$ (system operating normally), $j = 1$ (system degraded due to human error made by maintenance personnel), $j = 2$ (system failed).
$P_j(t)$ is the probability that the system is in state j at time t, for $j = 0, 1, 2$.

Using the Markov method and Fig. 11.4, we write down the following equations [1, 7, 8]:

$$\frac{dP_0(t)}{dt} + (\lambda_1 + \lambda) P_0(t) = \mu_1 P_1(t) + \mu P_2(t) \tag{11.45}$$

$$\frac{dP_1(t)}{dt} + (\mu_1 + \lambda_2) P_1(t) = \mu_2 P_2(t) + P_0(t)\lambda_1 \tag{11.46}$$

$$\frac{dP_2(t)}{dt} + (\mu + \mu_2) P_2(t) = \lambda_2 P_1(t) + \lambda P_0(t) \tag{11.47}$$

At time $t = 0$, $P_0(0) = 1$, $P_1(0) = 0$, and $P_2(0) = 0$.

By solving Equations (11.45)–(11.47), we get

$$
\begin{aligned}
P_0(t) = & \frac{\mu_1\mu + \lambda_2\mu + \mu_1\mu_2}{A_1 A_2} \\
& + \left[\mu_1 A_1 + \mu A_1 + \mu_2 A_1 + \lambda_2 A_1 + A_1^2 + \mu_1\mu + \lambda_2\mu + \mu_1\mu_2 \right] e^{A_1 t} \\
& + \left\{ 1 - \left(\frac{\mu_1\mu + \lambda_2\mu + \mu_1\mu_2}{A_1 A_2} \right) \right. \\
& \left. - \left[\frac{\mu_1 A_1 + \mu A_1 + \mu_2 A_1 + A_1\lambda_2 + A_1^2 + \mu_1\mu + \lambda_2\mu + \mu_1\mu_2}{A_1(A_1 - A_2)} \right] \right\} e^{A_2 t}
\end{aligned}
\tag{11.48}
$$

where

$$A_1, A_2 = \frac{\left[-D \pm \sqrt{D^2 - 4(\mu_1\mu + \lambda_2\mu + \mu_1\mu_2 + \mu\lambda_1 + \lambda_1\mu_2 + \lambda_1\lambda_2 + \mu_1\lambda + \lambda\mu_2 + \lambda\lambda_2)}\right]}{2}$$

$$D = \lambda_1 + \lambda + \lambda_2 + \mu_1 + \mu + \mu_2$$

$$A_1 A_2 = \mu_1\mu + \lambda_2\mu + \mu_1\mu_2 + \mu\lambda_1 + \lambda_1\mu_2 + \lambda_1\lambda_2 + \mu_1\lambda + \lambda\mu_2 + \lambda\lambda_2$$

$$
\begin{aligned}
P_1(t) = & \frac{\lambda_1\mu + \lambda_1\mu_2 + \lambda\mu_2}{A_1 A_2} + \left[\frac{A_1\lambda_1 + \lambda_1\mu + \lambda_1\mu_2 + \lambda\mu_2}{A_1(A_1 - A_2)} \right] e^{A_1 t} \\
& - \left[\frac{\lambda_1\mu + \lambda_1\mu_2 + \lambda\mu_2}{A_1 A_2} + \frac{A_1\lambda_1 + \lambda_1\mu + \lambda_1\mu_2 + \lambda\mu_2}{A_1(A_1 - A_2)} \right] e^{A_2 t}
\end{aligned}
\tag{11.49}
$$

$$
\begin{aligned}
P_2(t) = & \frac{\lambda_1\lambda_2 + \mu_1\lambda + \lambda\lambda_2}{A_1 A_2} + \left[\frac{A_1\lambda + \lambda_1\lambda_2 + \lambda\mu_1 + \lambda\lambda_2}{A_1(A_1 - A_2)} \right] e^{A_1 t} \\
& - \left[\frac{\lambda_1\lambda_2 + \mu_1\lambda + \lambda\lambda_2}{A_1 A_2} + \frac{A_1\lambda + \lambda_1\lambda_2 + \mu_1\lambda + \lambda\lambda_2}{A_1(A_1 - A_2)} \right] e^{A_2 t}
\end{aligned}
\tag{11.50}
$$

The probability of system degradation due to human error by maintenance personnel is given by Equation (11.49). As time t becomes very large, Equation (11.49) reduces to

$$P_1 = \frac{\lambda_1 \mu + \lambda_1 \mu_2 + \lambda \mu_2}{A_1 A_2} \tag{11.51}$$

where P_1 is the steady-state probability of system degradation due to human error by maintenance personnel.

The time-dependent system operational availability is given by

$$AV_S(t) = P_0(t) + P_1(t) \tag{11.52}$$

where $AV_S(t)$ is the system operational availability at time t.

As t becomes very large, Equation (11.52) becomes

$$AV_S = \frac{\mu_1 \mu + \lambda_2 \mu + \mu_1 \mu_2 + \lambda_1 \mu + \lambda_1 \mu_2 + \lambda \mu_2}{A_1 A_2} \tag{11.53}$$

where AV_S is the system steady-state operational availability.

EXAMPLE 11.5

Assume that for a system we have the following data values:

$$\lambda = 0.007 \text{ failures per hour}$$
$$\lambda_1 = 0.0002 \text{ errors per hour}$$
$$\lambda_2 = 0.002 \text{ failures per hour}$$
$$\mu = 0.03 \text{ repairs per hour}$$
$$\mu_1 = 0.006 \text{ repairs per hour}$$
$$\mu_2 = 0.04 \text{ repairs per hour}$$

Calculate the steady-state probability of system degradation due to human error by maintenance personnel.

By inserting the specified data values into Equation (11.51), we obtain

$$P_1 = \frac{(0.0002)(0.03) + (0.0002)(0.04) + (0.007)(0.04)}{(0.006)(0.03) + (0.002)(0.03) + (0.006)(0.04) + (0.03)(0.0002) + (0.0002)(0.04)}$$

$$\frac{}{+ (0.0002)(0.002) + (0.006)(0.007) + (0.007)(0.04) + (0.007)(0.002)}$$

$$= 0.3540$$

Thus, the steady-state probability of system degradation due to human error by maintenance personnel is 0.3540.

11.4 MODELS FOR PERFORMING REDUNDANT SYSTEMS MAINTENANCE ERROR ANALYSIS

Past experiences indicate that human error by maintenance personnel can cause not only the failure of single unit systems but also of redundant unit systems. In the published literature, there are many mathematical models that can be used to perform maintenance error analysis of redundant systems [1]. Two of these models are presented below.

11.4.1 MODEL I

This mathematical model represents a two-identical-units parallel system subjected to periodic preventive maintenance. The system/unit can fail due to hardware failures or maintenance or other errors. The system state space diagram is shown in Figure 11.5. The numerals in circles and boxes denote system states.

The following assumptions are associated with the model:

- All failures and errors occur independently.
- Both units are independent, active, and identical.
- Maintenance or other errors may occur when either both system units are good or when one system unit is good.
- The system is subjected to periodic preventive maintenance.
- Both failure and error rates are constant.
- The total system fails due to maintenance or other errors.

The following symbols are associated with the diagram:

i is the ith state of the system; $i = 0$ (both units operating normally), $i = 1$ (one unit failed due to hardware failure, the other operating normally), $i = 2$ (system failed due to maintenance or other errors), $i = 3$ (system failed due to hardware failures).

$P_i(t)$ is the probability that the system is in state i at time t, for $i = 0, 1, 2, 3$.

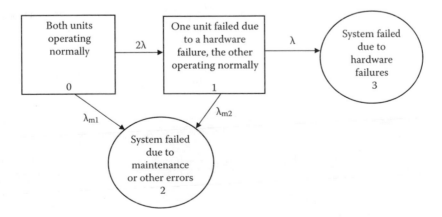

FIGURE 11.5 State space diagram for model I.

λ is the unit constant failure rate.

λ_{m1} is the constant maintenance or other error rate when both units are operating normally.

λ_{m2} is the constant maintenance or other error rate when only one unit is operating normally.

Using the Markov method and Figure 11.5, we get the following equations [1, 8]:

$$\frac{dP_0(t)}{dt} + (2\lambda + \lambda_{m1})\, P_0(t) = 0 \tag{11.54}$$

$$\frac{dP_1(t)}{dt} + (\lambda + \lambda_{m2})\, P_1(t) = 2\lambda P_0(t) \tag{11.55}$$

$$\frac{dP_2(t)}{dt} = \lambda_{m1} P_0(t) + \lambda_{m2}\, P_1(t) \tag{11.56}$$

$$\frac{dP_3(t)}{dt} = \lambda P_1(t) \tag{11.57}$$

At time $t = 0$, $P_0(0) = 1$, $P_1(0) = 0$, $P_2(0) = 0$, and $P_3(0) = 0$.
By solving Equations (11.54)–(11.57), we obtain

$$P_0(t) = e^{-A_1 t} \tag{11.58}$$

where

$$A_1 = 2\lambda + \lambda_{m1} \tag{11.59}$$

$$P_1(t) = B_1(e^{-A_1 t} - e^{-A_2 t}) \tag{11.60}$$

where

$$A_2 = \lambda + \lambda_{m2} \tag{11.61}$$

$$B_1 = \frac{2\lambda}{A_2 - A_1} \tag{11.62}$$

$$P_2(t) = B_2 - B_3 e^{-A_1 t} - B_4 e^{-A_2 t} \tag{11.63}$$

where

$$B_2 = \frac{2\lambda\lambda_{m2} + \lambda_{m1}A_2}{A_1 A_2} \tag{11.64}$$

$$B_3 = \frac{2\lambda\lambda_{m2} + \lambda_{m1}(A_2 - A_1)}{A_1(A_2 - A_1)} \tag{11.65}$$

$$B_4 = \frac{2\lambda\lambda_{m2}}{A_1(A_1 - A_2)} \tag{11.66}$$

$$P_3(t) = B_5 - B_6 e^{-A_1 t} - B_7 e^{-A_2 t} \tag{11.67}$$

where

$$B_5 = \frac{2\lambda^2}{A_1 A_2} \tag{11.68}$$

$$B_6 = \frac{2\lambda^2}{A_1(A_2 - A_1)} \tag{11.69}$$

$$B_7 = \frac{2\lambda^2}{A_2(A_1 - A_2)} \tag{11.70}$$

The system reliability is given by

$$\begin{aligned} R_S(t) &= P_0(t) + P_1(t) \\ &= e^{-A_1 t} + B_1(e^{-A_1 t} - e^{-A_2 t}) \end{aligned} \tag{11.71}$$

where $R_S(t)$ is the system reliability at time t.

The system mean time to failure is given by [1, 8]

$$\begin{aligned} MTTF_S &= \int_0^\infty R_S(t)dt \\ &= \int_0^\infty \left[e^{-A_1 t} + B_1(e^{-A_1 t} - e^{-A_2 t}) \right] dt \\ &= \frac{3\lambda + \lambda_{m2}}{(2\lambda + \lambda_{m1})(2\lambda + \lambda_{m2})} \end{aligned} \tag{11.72}$$

where $MTTF_S$ is the system mean time to failure.

EXAMPLE 11.6

Assume that a system is composed of two independent and identical units in parallel. The unit constant failure rate and the constant maintenance or other error rate when both units operate normally are 0.02 failures/hour and 0.004 errors/hour, respectively. The constant maintenance or other error rate, when only one unit operates normally, is 0.001 errors/hour.

Calculate the system mean time to failure.

By substituting the given data values into Equation (11.72), we get

$$MTTF_s = \frac{3(0.02) + 0.001}{[2(0.02) + 0.004](0.02 + 0.001)}$$

$$= 66.01 \text{ hours}$$

Thus, the system mean time to failure is 66.01 hours.

11.4.2 MODEL II

This model represents a system with two independent and identical units forming a parallel configuration subjected to periodic maintenance and failed unit repair. The system/unit can malfunction due to hardware failures or maintenance or other errors. The system state space diagram is shown in Figure 11.6. The numerals in boxes and circles denote system states.

The model is subjected to the following assumptions:

- Both units are active, independent, and identical.
- All failure, error, and repair rates are constant.
- All failures and errors occur independently.
- The total system fails due to maintenance or other errors.

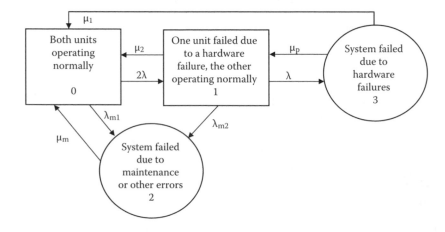

FIGURE 11.6 State space diagram for model II.

- Maintenance or other errors may occur when either both system units are good or when one system unit is good.
- The repaired system or unit is as good as new.

The following symbols are associated with Figure 11.6:

λ is the unit constant failure rate.
λ_{m1} is the constant maintenance or other error rate when both units are operating normally.
λ_{m2} is the constant maintenance or other error rate when only one unit is operating normally.
j is the jth state of the system; $j = 0$ (both units operating normally), $j = 1$ (one unit failed due to a hardware failure, the other operating normally), $j = 2$ (system failed due to maintenance or other errors), $j = 3$ (system failed due to hardware failures).
$P_j(t)$ is the probability that the system is in state j at time t, for $j = 0, 1, 2, 3$.
μ_1 is the system constant repair rate from state 3 to state 0.
μ_2 is the system constant repair rate from state 1 to state 0.
μ_m is the system constant repair rate from state 2 to state 0.
μ_p is the system constant repair rate from state 3 to state 1.

By using the Markov method and Figure 11.6, we write down the following equations [1, 8]:

$$\frac{dP_0(t)}{dt} + (2\lambda + \lambda_{m1}) P_0(t) = P_1(t) \mu_2 + P_3(t) \mu_1 + P_2(t) \mu_m \tag{11.73}$$

$$\frac{dP_1(t)}{dt} + (\lambda + \lambda_{m2} + \mu_2) P_1(t) = P_0(t) 2\lambda + P_3(t) \mu_p \tag{11.74}$$

$$\frac{dP_2(t)}{dt} + \mu_m P_2(t) = P_0(t) \lambda_{m1} + P_1(t) \mu_{m2} \tag{11.75}$$

$$\frac{dP_3(t)}{dt} + (\mu_p + \mu_1) P_3(t) = P_1(t) \lambda \tag{11.76}$$

At time $t = 0$, $P_0(0) = 1$, $P_1(0) = 0$, $P_2(0) = 0$, and $P_3(0) = 0$.
By solving Equations (11.73)–(11.76), we obtain the following steady-state probability equations [1, 8]:

$$P_0 = \left[1 + D_1 + 2\lambda^2 D + \frac{1}{\mu_m} (\lambda_{m1} + D_1 \lambda_{m2}) \right]^{-1} \tag{11.77}$$

where

$$D = [(\mu_1 + \mu_p)(\lambda + \lambda_{m2} + \mu_2) - \lambda\mu_p]^{-1}$$

$$D_1 = 2\lambda(1 + \lambda\mu_p D)/(\lambda + \lambda_{m2} + \mu_2)$$

$$P_1 = P_0 D_1 \tag{11.78}$$

$$P_2 = P_0(\lambda_{m1} + D_1\lambda_{m2})/\mu_m \tag{11.79}$$

$$P_3 = P_0 2\lambda^2 D \tag{11.80}$$

where P_0, P_1, P_2, and P_3 are the steady-state probabilities of the system being in states 0, 1, 2, and 3, respectively.

The system steady-state availability is given by

$$AV_{SS} = P_0 + P_1 \tag{11.81}$$

where AV_{SS} is the system steady-state availability.

Additional information on this model is available in Refs. [1, 9].

11.5   PROBLEMS

1. A maintenance worker is performing a certain task and his or her error rate is 0.004 errors/hour (i.e., times to human error are exponentially distributed). Calculate the maintenance worker's reliability during an 8-hour work period.
2. Prove Equation (11.28) by using Equation (11.27).
3. Assume that a maintenance worker's constant error rates in normal and stressful environments are 0.0002 errors/hour and 0.0006 errors/hour, respectively. The values of the transition rates from normal to stressful environment and vice versa are 0.004 times per hour and 0.006 times per hour, respectively. Calculate the mean time to human error of the maintenance worker.
4. Prove that the sum of Equations (11.32)–(11.34) is equal to unity and explain why.
5. A system can fail either due to human errors made by maintenance personnel or due to hardware failures. The system constant human error and hardware failure rates are 0.0002 errors/hour and 0.0008 failures/hour, respectively. Calculate the probability that the system will fail due to a human error made by maintenance personnel during a 10-hour mission.
6. Prove Equations (11.40)–(11.42) by using Equations (11.37)–(11.39).
7. Prove Equation (11.51) by using Equation (11.49).
8. Assume that for a system we have the following data values:

$\lambda = 0.008$ failures/hour
$\lambda_1 = 0.0001$ errors/hour
$\lambda_2 = 0.002$ failures/hour
$\mu = 0.02$ repairs/hour
$\mu_1 = 0.004$ repairs/hour
$\mu_2 = 0.03$ repairs/hour

Calculate the steady-state probability of system degradation due to human error by maintenance personnel, by using Equation (11.51).

9. A system is composed of two independent and identical units in parallel. The unit constant failure rate and the constant maintenance or other error rate when both units operate normally are 0.03 failures/hour and 0.005 errors/hour, respectively. The constant maintenance or other error rate when only one unit operates normally is 0.002 errors/hour. Calculate the system mean time to failure.

10. Prove Equations (11.77)–(11.80) by using Equations (11.73)–(11.76).

REFERENCES

1. Dhillon, B.S., *Human Reliability: With Human Factors*, Pergamon Press, New York, 1986.
2. Regulinski, T.L., Askren, W.B., Stochastic Modeling of Human Performance Effectiveness Functions, *Proceedings of the Annual Reliability and Maintainability Symposium*, 1972, pp. 407–416.
3. Dhillon, B.S., *Human Reliability and Error in Transportation Systems*, Springer, London, 2007.
4. Askren, W.B., Regulinski, T.L., Quantifying Human Performance for Reliability Analysis of Systems, *Human Factors*, Vol. 11, 1969, pp. 393–396.
5. Regulinski, T.L., Askren, W.B., Mathematical Modeling of Human Performance Reliability, *Proceedings of the Annual Symposium on Reliability*, 1969, pp. 5–11.
6. Dhillon, B.S., Stochastic Models for Predicting Human Reliability, Microelectronics and Reliability, Vol. 25, 1985, pp. 729–752.
7. Dhillon, B.S., *Design Reliability: Fundamentals and Applications*, CRC Press, Boca Raton, FL, 1999.
8. Dhillon, B.S., *Engineering Maintenance: A Modern Approach*, CRC Press, Boca Raton, FL, 2002.
9. Dhillon, B.S., Rayapati, S.N., Analysis of Redundant Systems with Human Errors, *Proceedings of the Annual Reliability and Maintainability Symposium*, 1985, pp. 315–321.

Appendix

A.1 INTRODUCTION

Over the years, a large number of publications on human reliability, error, and human factors in engineering maintenance have appeared in the form of journal articles, conference proceedings articles, technical reports, and so on. This appendix presents an extensive list of selective publications related, directly or indirectly, to human reliability, error, and human factors in engineering maintenance.

The period covered by the listing is from 1929 to 2007. The main objective of this listing is to provide readers with sources for obtaining additional information on human reliability, error, and human factors in engineering maintenance.

A.2 PUBLICATIONS

1. Adams, S.K., Sabri, Z.A., Husseiny, A.A., Maintenance and Testing Errors in Nuclear Power Plants: A Preliminary Assessment, *Proceedings of the Human Factors 24th Annual Meeting*, 1980, 280–284.
2. Agnihotri, R.K., Singhal, G., Khandelwal, S.K., Stochastic Analysis of a Two-Unit Redundant System with Two Types of Failure, *Microelectronics and Reliability*, Vol. 32, No. 7, 1992, pp. 901–904.
3. Allen, J.O. Rankin, W.L., Use of the Maintenance Error Decision Aid (MEDA) to Enhance Safety and Reliability and Reduce Costs in the Commercial Aviation Industry, *Proceedings of the Tenth Federal Aviation Administration Meeting on Human Factors Issues in Aircraft Maintenance and Inspection: Maintenance Performance Enhancement and Technician Resource Management*, 1996, pp. 79–87.
4. Allen, J.P., Marx, D.M., Maintenance Error Decision Aid Project, *Proceedings of the Eighth Federal Aviation Administration Meeting on Human Factors Issues in Aircraft Maintenance and Inspection: Trends and Advances in Aviation Maintenance Operations*, 1994, pp. 101–116.
5. Allen, J.P., Rankin, W.L., Summary of the Use and Impact of the Maintenance Error Decision Aid (MEDA) on the Commercial Aviation Industry, *Proceedings of the International Air Safety Seminar*, 1995, pp. 359–369.
6. Amalberti, R., Wioland, L., Human Error in Aviation, *Proceedings of the International Aviation Safety Conference on Aviation Safety: Human Factors, System Engineering, Flight Operations, Economics, and Strategies Management*. 1997, pp. 91–108.
7. Anderson, D.E., Malone, T.B., Baker, C.C., Recapitalizing the Navy through Optimized Manning and Improved Reliability, *Naval Engineers Journal*, November 1998, pp. 61–72.
8. Bacchi, M., Cacciabue, C., O'Connor, S., Reactive and Proactive Methods for Human Factors Studies in Aviation Maintenance, *Proceedings of the Ninth International Symposium on Aviation Psychology*, 1997, pp. 991–996.
9. Balkey, J.P., Human Factors Engineering in Risk-Based Inspections, *Safety Engineering and Risk Analysis*, Vol. 6, 1996, pp. 97–106.
10. Banker, R.D., Datar, S.M., Kemerer, C.F., Software Errors and Software Maintenance Management, *Information Technology and Management*, Vol. 3, No. 1–2, 2002, 25–41.

11. Benham, H., Spreadsheet Structure and Maintenance Errors, *Proceedings of the Information Resources Management Association International Conference*, 1993, pp. 262–267.

12. Blackmon, R.B., Gramopadhye, A.K., Using the Aircraft Inspector's Training System to Improve the Quality of Aircraft Inspection, *Proceedings of the 5th Industrial Engineering Research Conference*, 1996, pp. 447–452.

13. Blenkinsop, G., Only Human, *Quality World*, Vol. 29, No. 12, 2003, pp. 24–29.

14. Bos, T., Hoekstra, R., Reduction of Error Potential in Aircraft Maintenance, Available from the Human Factors Department, National Aerospace Laboratory NLR, P.O. Box 90502, 1006 BM Amsterdam, The Netherlands.

15. Bowen, J.B., Software Maintenance: An Error-Prone Activity, *Proceedings of the IEEE Software Maintenance Workshop*, 1984, pp. 102–105.

16. Brackenbury, K.F., The Human Factor in Maintenance, *Motor Transport*, Vol. 48, No. 1243–1245, Jan. 1929, pp. 5–6, 53–54, and 81–82.

17. Bradley, E.A., Case Studies in Disaster—A Coded Approach, *International Journal of Pressure Vessels & Piping*, Vol. 61, No. 2–3, 1995, pp. 177–197.

18. Braithwaite, G.R., A Safe Culture or Safety Culture, *Proceedings of the Ninth International Symposium on Aviation Psychology*, 1997, pp. 1029–1031.

19. Carr, M.J., Christer, A.H., Incorporating the Potential for Human Error in Maintenance Models, *Journal of the Operational Research Society*, Vol. 54, No. 12, 2003, pp. 1249–1253.

20. Cavalier, M.P., Knapp, G.M., Reducing Preventive Maintenance Cost Error Caused by Uncertainty, *Journal of Quality in Maintenance Engineering*, Vol. 2, No. 3, 1996, pp. 21–36.

21. Celeux, G., Corset, F., Garnero, M.A., Breuils, C., Accounting for Inspection Errors and Change in Maintenance Behaviour, *IMA Journal of Management Mathematics*, Vol. 13, No. 1, 2002, pp. 51–59.

22. Chandler, T.N., Reducing Federal Aviation Regulation (FAR) Errors through STAR, *Proceedings of the 11th FAA/AAM Meeting on Human Factors in Aviation Maintenance and Inspection*, 1997, pp. 108–118.

23. Cheng-Yi, L., Optimization of Maintenance, Production and Inspection Strategies While Considering Preventive Maintenance Error, *Journal of Information and Optimization Sciences*, Vol. 25, No. 3, 2004, pp. 543–555.

24. Chervak, S.G., Drury, C.G., Human Factors Audit Program for Maintenance, in Human Factors in Aviation Maintenance–Phase V: Progress Report, Office of Aviation Medicine, Federal Aviation Administration, Washington, D.C., 1995, pp. 93–126.

25. Chien-Kuo, S., Cheng-Yi, L., Optimizing an Integrated Production and Quality Strategy Considering Inspection and Preventive Maintenance Errors, *Journal of Information and Optimization Sciences*, Vol. 27, No. 3, 2006, pp. 577–593.

26. Christensen, J.M., Human Factors in Maintenance, *Proceedings of the Conference on Pressure Vessel Piping Technology*, 1985, pp. 721–731.

27. Chung, W.K., Reliability Analysis of a Repairable Parallel System with Standby Involving Human Error and Common-Cause Failure, *Microelectronics and Reliability*, Vol. 27, No. 2, 1987, pp. 269–271.

28. Ciavarelli, A.J., Organizational Factors in Aviation Accidents, *Proceedings of the Ninth International Symposium on Aviation Psychology*, April 27–May 1, 1997, pp. 310–316.

29. Cunningham, B.G., Maintenance Human Factors at Northwest Airlines, *Proceedings of the 10th Federal Aviation Administration, Meeting on Human Factors Issues in Aircraft Maintenance and Inspection: Maintenance Performance Enhancement and Technician Resource Management*, 1996, pp. 43–53.

30. Danahar, J.W., Maintenance and Inspection Issues in Aircraft Accidents/Incidents, *Proceedings of the Meeting on Human Factors Issues in Aircraft Maintenance and Inspection*, 1989, pp. A9–A11.
31. Daniels, R.W., The Formula for Improved Plant Maintainability Must Include Human Factors, *Proceedings of the IEEE 3rd Conference on Human Factors and Power Plants*, 1985, pp. 242–244.
32. Darwish, M.A., Ben-Daya, M., Effect of Inspection Errors and Preventive Maintenance on a Two-Stage Production Inventory System, *International Journal of Production Economics*, Vol. 107, No. 1, 2007, pp. 301–303.
33. Desormiere, D., Impact of New Generation Aircraft on the Maintenance Environment and Work Procedures, *Proceedings of the 5th Federal Aviation Administration Meeting on Human Factors Issues in Aircraft Maintenance and Inspection: The Work Environment in Aviation Maintenance*, 1992, pp. 124–134.
34. Dhillon, B.S., *Human Reliability: With Human Factors,* Pergamon Press, New York, 1986.
35. Dhillon, B.S., Modeling Human Errors in Repairable Systems, *Proceedings of the Annual Reliability and Maintainability Symposium*, 1989, pp. 418–424.
36. Dhillon, B.S., *Engineering Maintenance: A Modern Approach*, CRC Press, Boca Raton, Florida, 2002.
37. Dhillon, B.S., Liu, Y., Human Error in Maintenance: A Review, *Journal of Quality in Maintenance Engineering*, Vol. 12, No. 1, 2006, pp. 21–36.
38. Dhillon, B.S., Rayapati, S.N., Human Error and Common-Cause Failure Modelling of Standby Systems, *Maintenance Management International*, Vol. 7, No. 2, 1988, pp. 93–110.
39. Dhillon, B.S., Rayapati, S.N., Human Error Modelling of Parallel and Standby Redundant Systems, *International Journal of Systems Science*, Vol. 19, No. 4, 1988, pp. 589–611.
40. Dhillon, B.S., Yang, N., Probabilistic Analysis of a Maintainable System with Human Error, *Journal of Quality in Maintenance Engineering*, Vol. 1, No. 2, 1995, pp. 50–59.
41. Dhillon, B.S., Yang, N.F., Human Error Analysis of a Standby Redundant System with Arbitrarily Distributed Repair Times, *Microelectronics and Reliability*, Vol. 33, No. 3, 1993, pp. 431–444.
42. Dhillon, B.S., Yang, N.F., Probabilistic Analysis of a Maintainable System with Human Error, *Journal of Quality in Maintenance Engineering*, Vol. 1, No. 2, 1995, pp. 50–59.
43. Doll, R., Maintenance and Inspection Issues in Air Carrier Operations, In Human Factors Issues in Aircraft Maintenance and Inspection, Report No. DOT/FAA/AAM-89/9, Office of the Aviation Medicine, Federal Aviation Administration, Washington, D.C., 1989, pp. A33–36.
44. Dorn, M.D., Effects of Maintenance Human Factors in Maintenance-Related Aircraft Accidents, *Transportation Research Record*, No. 1517, 1996, pp. 17–28.
45. Drury, C.G., Errors in Aviation Maintenance: Taxonomy and Control, *Proceedings of the 35th Annual Meeting of the Human Factors Society*, 1991, pp. 42–46.
46. Drury, C.G., Murthy, M.R., Wenner, C.L., A Proactive Error Reduction System, *Proceedings of the 11th Federal Aviation Administration Meeting on Human Factors Issues in Aircraft Maintenance and Inspection: Human Error in Aviation Maintenance*, 1997, pp. 91–103.
47. Drury, C.G., Rangel, J., Reducing Automation–Related Errors in Maintenance and Inspection, In Human Factors in Aviation Maintenance–Phase VI: Progress Report, Vol. II, Office of Aviation Medicine, Federal Aviation Administration, Washington, D.C., 1996, pp. 281–306.

48. Drury, C.G., Shepherd, W.T., Johnson, W.B., Error Reduction in Aviation Maintenance, *Proceedings of the 13th Triennial Congress of the International Ergonomics Association*, 1997, pp. 31–33.

49. Drury, C.G., Spencer, F.W., Measuring Human Reliability in Aircraft Inspection, *Proceedings of the 13th Triennial Congress of the International Ergonomics Association*, 1997, pp, 34–36.

50. Drury, C.G., Wenner, C.L., Murthy, M., A Proactive Error Reduction System, *Proceedings of the 11th FAA/AAM Meeting on Human Factors in Aviation Maintenance and Inspection*, 1997, pp. 93–106.

51 Drury, C.G., Wenner, C.L., Murthy, M., A Proactive Error Reporting System, In Human Factors in Aviation Maintenance – Phase VII: Progress Report, Office of Aviation Medicine, Federal Aviation Administration, Washington, D.C., 1997, pp. 173–184.

52. Dunn, S., Managing Human Error in Maintenance, Available from Assetivity Pty Ltd., P.O. Box 1315, Booragoon, WA 6154, U.S.A.

53. Duphily, R.J., Human Factors Considerations during Operations and Maintenance, *Proceedings of the IEEE Global Telecommunications Conference and Exhibition*, 1989, pp. 792–794.

54. DuPont, G., The Dirty Dozen Errors in Maintenance, *Proceedings of the 11th FAA/AAM Meeting on Human Factors in Aviation Maintenance and Inspection*, 1997, pp. 49–52.

55. Eiff, G.M., Lopp, D., Abdul, Z., Lapacek, M., Ropp, T., Practical Considerations of Maintenance Human Factors for Line Operations, *Proceedings of the 11th FAA/AAM Meeting on Human Factors in Aviation Maintenance and Inspection*, 1997, pp. 125–139.

56. Emling, J.W., Human Factors in Transmission Maintenance, *Bell Laboratories Record*, Vol. 40, No. 4, 1962, pp. 130–136.

57. Endsley, M.R., Robertson, M.M., Team Situation Awareness in Aviation Maintenance, *Proceedings of the 10th Federal Aviation Administration Meeting on Human Factors Issues in Aircraft Maintenance and Inspection: Maintenance Performance Enhancement and Technician Resource Management*, 1996, pp. 95–101.

58. Endsley, M.R., Robertson, M.M., Situation Awareness in Aircraft Maintenance Teams, *International Journal of Industrial Ergonomics*, Vol. 26, 2000, pp. 301–325.

59. Eves, D.C.T., *Deadly Maintenance: A Study of Fatal Accidents at Work*, Her Majesty's Stationery Office (HMSO), London, UK, 1985.

60. Fitzpatrik, J., Wright, M., Qantas Engineering and Maintenance Human Factors: The Human Error and Accident Reduction (HEAR) Programme, *Proceedings of the 11th FAA/AAM Meeting on Human Factors in Aviation Maintenance and Inspection*, 1997, pp. 25–34.

61. Ford, T., Three Aspects of Aerospace Safety: Human Factors in Airline Maintenance, Aircraft Engineering and Aerospace Technology, Vol. 69, No. 3, 1997, pp. 262–264.

62. Fotos, C.P., Continental Applies CRM Concepts to Technical, Maintenance Corps, *Aviation Week and Space Technology*, August 26, 1991, pp. 32–35.

63. Gertman, D.I., Conversion of a Mainframe Simulation for Maintenance Performance to a PC Environment, *Reliability Engineering & System Safety*, Vol. 38, No. 3, 1992, pp. 211–217.

64. Goglia, J., Maintenance Training: A Review from the Floor, *Proceedings of the 3rd Federal Aviation Administration Meeting on Human Factors Issues in Aircraft Maintenance and Inspection: Training Issues*, 1990, pp. 79–82.

65. Graeber, R.C., Marx, D.A., Reducing Human Error in Aircraft Maintenance Operations, *Proceeding of the 46th Annual International Air Safety Seminar & International Federation of Airworthiness 23rd International Conference*, November 8–11, 1993, pp. 147–157.

66. Graham, D.B., Kuenzi, J.K., Error Control Systems at Northwest Airlines, *Proceedings of the 11th FAA/AAM Meeting on Human Factors in Aviation Maintenance and Inspection*, 1997, pp. 35–41.

67. Gramopadhye, A.K., Drury, C.G., Human Factors in Aviation Maintenance: How We Get to Where We Are? *International Journal of Industrial Ergonomics*, Vol. 26, No. 2, 2000, pp. 125–131.

68. Gramopadhye, A.K., Drury, C.G., Prabhu, P., Training for Aircraft Visual Inspection, *Human Factors and Ergonomics in Manufacturing*, Vol. 3, 1997, pp. 171–196.

69. Grubb, N.S., Inspection and Maintenance Issues in Commuter Air Carrier Operations, *Proceedings of the Meeting on Human Factors Issues in Aircraft Maintenance and Inspection*, 1989, pp. A37–A41.

70. Gupta, P.P., Sharma, R.K., Reliability Analysis of a Two State Repairable Parallel Redundant System Under Human Failure, *Microelectronics and Reliability*, Vol. 26, No. 2, 1986, pp. 221–224.

71. Gupta, P.P., Singhal, A., Singh, S.P., Cost Analysis of a Multi-Component Parallel Redundant Complex System with Overloading Effect and Waiting under Critical Human Error, *Microelectronics and Reliability*, Vol. 31, No. 5, 1991, pp. 865–868.

72. Hal, D., The Role of Human Factors Training and Error Management in the Aviation Maintenance Safety System, *Proceedings of the Flight Safety Foundation Annual International Air Safety Seminar*, 2005, pp. 245–249.

73. Havard, S., Why Adopt a Human Factors Program in Engineering? *Proceedings of the Third Australian Aviation Psychology Symposium*, 1995, pp. 394–399.

74. Hibit, R., Marx, D.A., Reducing Human Error in Aircraft Maintenance Operations with the Maintenance Error Decision Aid (MEDA), *Proceedings of the Human Factors and Ergonomics Society 38th Annual Meeting*, 1994, pp. 111–114.

75. Hobbs, A., Maintenance Mistakes and System Solutions, *Asia Pacific Air Safety*, Vol. 21, 1999, pp. 1–7.

76. Hobbs, A., Robertson, M.M., Human Factors in Aircraft Maintenance Workshop Report, *Proceedings of the Third Australian Aviation Psychology Symposium*, 1995, pp. 468–474.

77. Hobbs, A., Williamson, A., Human Factors in Airline Maintenance, *Proceedings of the Australian Aviation Psychology Symposium*, 1995, pp. 384–393.

78. Hobbs, A., Williamson, A., Skills, Rules and Knowledge in Aircraft Maintenance: Errors in Context, *Ergonomics*, Vol. 45, No. 4, 2002, pp. 290–308.

79. Hobbs, A., Williamson, A., Associations between Errors and Contributing Factors in Aircraft Maintenance, *Human Factors*, Vol. 42, No. 2, 2003, pp. 186–201.

80. Huang, W.G., Zhang, L., Cause Analysis and Preventives for Human Error Events in Daya Bay NPP, *Dongli Gongcheng/Nuclear Power Engineering*, Vol. 19, No. 1, 1998, pp. 64–67, 76.

81. Isoda, H., Yasutake, J.Y., Human Factors Interventions to Reduce Human Errors and Improve Productivity in Maintenance Tasks, *Proceedings of the International Conference on Design and Safety of Advanced Nuclear Power Plants*, 1992, pp. 34.4/1–6.

82. Ivaturi, S., Gramopadhye, A.K., Kraus, D., Blackmon, R., Team Training to Improve the Effectiveness of Teams in the Aircraft Maintenance Environment, *Proceedings of the Human Factors and Ergonomics Society 39th Annual Meeting*, 1995, pp. 1355–1359.

83. Jacob, M., Narmada, S., Varghese, T., Analysis of a Two Unit Deteriorating Standby System with Repair, *Microelectronics and Reliability*, Vol. 37, No. 5, 1997, pp. 857–861.

84. Jacobsson, L., Svensson, O., Psychosocial Work Strain of Maintenance Personnel during Annual Outage and Normal Operation in a Nuclear Power Plant, *Proceedings of the Human Factors Society 35th Annual Meeting*, Vol. 2, 1991, pp. 913–917.

85. Johnson, W.B., National Plan for Aviation Human Factors: Maintenance Research Issues, *Proceedings of the Human Factors Society Annual Meeting*, 1991, pp. 28–32.

86. Johnson, W.B., Human Factors in Maintenance: An Emerging System Requirement, *Ground Effects*, Vol. 2, 1997, pp. 6–8.

87. Johnson, W.B., Norton, J.E., Using Intelligent Simulation to Enhance Human Performance in Aircraft Maintenance, *Proceedings of the International Conference on Aging Aircraft and Structural Airworthiness*, 1991, pp. 305–311.

88. Johnson, W.B, Rouse, W.B., Analysis and Classification of Human Error in Troubleshooting Live Aircraft Power Plants, *IEEE Transactions on Systems, Man, and Cybernetics*, Vol. 12, No. 3, 1982, pp. 389–393.

89. Johnson, W.B., Shepherd, W.T., Human Factors in Aviation Maintenance: Research and Development in the USA, *Proceedings of the ICAO Flight Safety and Human Factors Seminar*, 1991, pp. B.192–B.228.

90. Johnson, W.B., Shepherd, W.T., The Impact of Human Factors Research on Commercial Aircraft Maintenance and Inspection, *Proceedings of the Flight Safety Foundation 46th Annual International Air Safety Seminar*, 1993, pp. 187–200.

91. Jones, J.A., Widjaja, T.K., Electronic Human Factors Guide for Aviation Maintenance, *Proceedings of the Human Factors and Ergonomics Society Annual Meeting*, 1995, pp. 71–74.

92. Joong, N.K., The Development of K-HPES: A Korean-Version Human Performance Enhancement System, *Proceedings of IEEE Sixth Annual Human Factors Meeting*, 1997, pp. 1/16–1/20.

93. Kania, J., Panel Presentation on Airline Maintenance Human Factors, *Proceedings of the 10th FAA Meeting on Human Factors in Aircraft, FAA/AAM Human Factors in Aviation Maintenance and Inspection Research Phase Reports (1991–1999)*, Brussels, Belgium, 1997.

94. Kanki, B., Managing Procedural Error in Maintenance, *Proceedings of the Flight Safety Foundation Annual International Air Safety Seminar*, 2005, pp. 233–244.

95. Kanki, B.G., Walter, D., Reduction of Maintenance Error Through Focused Interventions, *Proceedings of the 11th FAA/AAM Meeting on Human Factors in Aviation Maintenance and Inspection*, 1997, pp. 120–124.

96. Kirby, M.J., Klein, R.L., Separation of Maintenance and Operator Errors from Equipment Failures, *Proceedings of the Product Assurance Conference and Technical Exhibit*, 1969, pp. 17–27.

97. Klein, R., The Human Factors Impact of an Export System Based Reliability Centered Maintenance Program, *Proceedings of the IEEE Conference on Human Factors and Power Plants*, 1992, pp. 241–245.

98. Knee, H.E., The Maintenance Personnel Performance Simulation (MAPPS) Model: A Human Reliability Analysis Tool, *Proceedings of the International Conference on Nuclear Power Plant Aging, Availability Factor and Reliability Analysis*, 1985, pp. 77–80.

99. Koli, S., Chervak, S., Drury, C.G., Human Factors Audit Programs for Nonrepetitive Tasks, *Human Factors and Ergonomics in Manufacturing*, Vol. 8, No. 3, 1998, pp. 215–231.

100. Komarniski, R., Maintenance Human Factors and the Organization, *Proceedings of the Corporate Aviation Safety Seminar*, 1999, pp. 265–267.
101. Kraus, D., Gramopadhye, A.K, Role of Computers in Team Training: The Aircraft Maintenance Environment Example, *Proceedings of the 11th FAA/ AAM Meeting on Human Factors in Aviation Maintenance and Inspection*, 1997, pp. 54–57.
102. Kuo-Wei, S., Shueue-Ling, H., Thu-Hua, L., Knowledge Architecture and Framework Design for Preventing Human Error in Maintenance Tasks, *Expert Systems and Applications*, Vol. 19, No. 3, 2000, pp. 219–228.
103. Lafaro, R.J., Maintenance Resource Management: It Can't Be Crew Resource Management Re-packaged, *Proceedings of the 11th FAA/AAM Meeting on Human Factors in Aviation Maintenance and Inspection*, 1997, pp. 68–81.
104. Latorella, K.A., Drury, C.G., Human Reliability in Aircraft Inspection, In Human Factors in Aviation Maintenance Phase II: Progress Report, Report No. DOT/FAA/ AM-93/5, Office of Aviation Medicine, Federal Aviation Administration, Washington, D.C., 1993, pp. 63–144.
105. Latorella, K.A., Drury, C.G.A., Framework for Human Reliability in Aircraft Inspection, *Proceedings of the 7th Federal Aviation Administration Meeting on Human Factors Issues in Aircraft Maintenance and Inspection: Science, Technology, and Management: A Program Review*, 1992, pp. 71–82.
106. Latorella, K.A., Prabhu, P.V., Review of Human Error in Aviation Maintenance and Inspection, *International Journal of Industrial Ergonomics*, Vol. 26, No. 2, 2000, pp. 133–161.
107. Layton, C.F., Shepherd, W.T., Johnson, W.B., Human Factors and Aircraft Maintenance, *Proceedings of the International Air Transport Association 22nd Technical Conference on Human Factors in Maintenance*, 1993, pp. 143–154.
108. Layton, C.F, Shepherd, W.T., Johnson, W.B., Norton, J.E., Enhancing Human Reliability with Integrated Information Systems for Aviation Maintenance, *Proceedings of the Annual Reliability and Maintainability Symposium*, 1993, pp. 498–502.
109. Lee, J.W., Oh, I.S., Lee, H.C., Lee, Y.H., Sim, B.S., Human Factors Research in KAERI for Nuclear Power Plants, *Proceedings of the IEEE Sixth Annual Human Factors Meeting*, 1997, pp. 13/11–13/16.
110. Lee, S.C., Lee, E.T., Wang, Y.M., A New Scientific Accuracy Measure for Performance Evaluation of Human-Computer Diagnostic Systems, *Proceedings of SPIE—The International Society for Optical Engineering Conference*, Vol. 4553, 2001, pp. 203–214.
111. Maddox, M.E., Introducing a Practical Human Factors Guide into the Aviation Maintenance Environment, *Proceedings of the Human Factors and Ergonomics Society 38th Annual Meeting*, 1994, pp. 101–105.
112. Maddox, M.E., Providing Useful Human Factors Guidance to Aviation Maintenance Practitioners, *Proceedings of the Human Factors and Ergonomics Society 39th Annual Meeting*, 1995, pp. 66–70.
113. Maillart, L.M., Pollock, S.M., The Effect of Failure-Distribution Specification-Errors on Maintenance Costs, *Proceedings of the Annual Reliability and Maintainability Symposium*, 1999, pp. 69–77.
114. Maintenance Error Decision Aid, Boeing Commercial Airplane Group, Seattle, Washington, 1994.
115. Majoros, A.E., Human Performance in Aircraft Maintenance: The Role of Aircraft Design, *Proceedings of the Meeting on Human Factors Issues in Aircraft Maintenance and Inspection*, 1989, pp. A25–A32.

116. Majoros, A.E., Human Factors Issues in Manufacturers' Maintenance-Related Communication, *Proceedings of the 2nd Federal Aviation Administration Meeting on Human Factors Issues in Aircraft Maintenance and Inspection: Information Exchange and Communications*, 1990, pp. 59–68.

117. Majoros, A.E., Aircraft Design for Maintainability with Future Human Models, *Proceedings of the 6th Federal Aviation Administration Meeting on Human Factors Issues in Aircraft Maintenance and Inspection: Maintenance 2000*, 1992, pp. 49–63.

118. Manwaring, J.C., Conway, G.A., Garrett, L.C., Epidemiology and Prevention of Helicopter External Load Accidents, *Journal of Safety Research,* Vol. 29, No. 2, 1998, pp. 107–121.

119. Marksteiner, J.P., Maintenance, How Much Is Too Much? *Proceeding of the 52nd Annual International Air Safety Seminar (IASS)*, 1999, pp. 85–92.

120. Marx, D.A., Moving Toward 100% Error Reporting in Maintenance, *Proceedings of the 11th FAA/AAM Meeting on Human Factors in Aviation Maintenance and Inspection*, 1997, pp. 42–48.

121. Mason, S., Measuring Attitudes to Improve Electricians' Safety, *Mining Technology,* Vol. 78, No. 898, 1996, pp. 166–170.

122. Mason, S., Improving Maintenance by Reducing Human Error, 2007. Available from Health, Safety, and Engineering Consultant Ltd., 70 Tamworth Road, Ashby-de-la-Zouch, Leicestershire, UK.

123. Masson, M., Koning, Y., How to Manage Human Error in Aviation Maintenance? The Example of a Jar 66-HF Education and Training Programme, *Cognition, Technology & Work*, Vol. 3, No. 4, 2001, pp. 189–204.

124. McDonald, N., *Human-Centered Management Guide for Aircraft Maintenance*, Trinity College Press, Dublin, 2000.

125. McGrath, R.N., Safety and Maintenance Management: A View from an Ivory Tower, *Proceedings of the Aviation Safety Conference and Exposition*, 1999, pp. 21–26.

126. McRoy, S., Preface: Detecting, Repairing and Preventing Human-Machine Miscommunication, *International Journal of Human-Computer Studies*, Vol. 48, 1998, pp. 547–552.

127. McWilliams, T.P., Martz, H.F., Human Error Considerations in Determining the Optimum Test Interval for Periodically Inspected Standby Systems, *IEEE Transactions on Reliability*, Vol. 29, No. 4, 1980, pp. 305–310.

128. Meelot, M., Human Factor in Maintenance Activities in Operation, *Proceedings of the 10th International Conference on Power Stations*, 1989, 82.1–82.4.

129. Meghashyam, G., Electronic Ergonomic Audit System for Maintenance and Inspection, *Proceedings of the Human Factors and Ergonomics Society 39th Annual Meeting*, 1995, pp. 75–78.

130. Moran, J.T., Human Factors in Aircraft Maintenance and Inspection, Rotorcraft Maintenance and Inspection, *Proceedings of the Meeting on Human Factors Issues in Aircraft Maintenance and Inspection*, 1989, pp. A42–A44.

131. Morgan, C.B., Implementing Training Programs—Operation, Maintenance and Safety, *Proceedings of the 30th IEEE Cement Industry Technical Conference*, 1988, pp. 233–247.

132. Morgenstern, M.H., Maintenance Management Systems: A Human Factors Issue, *Proceedings of the IEEE Conference on Human Factors and Power Plants*, 1988, pp. 390–393.

133. Mount, F.E., Human Factor in Aerospace Maintenance, *Aerospace America*, Vol. 31, No. 10, 1993, pp. 1–9.

134. Mulzoff, M.T., Information Needs of Aircraft Inspectors, *Proceedings of the 2nd Federal Aviation Administration Meeting on Human Factors Issues in Aircraft Maintenance and Inspection: Information Exchange and Communications*, 1990, pp. 79–84.

135. Nakashima, T., Oyama, M., Hisada, H., Ishii, N., Analysis of Software Bug Causes and Its Prevention, *Information and Software Technology*, Vol. 41, 1999, pp. 1059–1068.

136. Nakatani, Y., Nakagawa, T., Terashita, N., Umeda, Y., Human Interface Evaluation by Simulation, *Proceedings of the IEEE Sixth Annual Human Factors Meeting*, 1997, pp. 7/18–7/23.

137. Narmada, S., Jacob, M., Reliability Analysis of a Complex System with a Deteriorating Standby Unit under Common-Cause Failure and Critical Human Error, *Microelectronics and Reliability*, Vol. 36, No. 9, 1996, pp. 1287–1290.

138. Nelson, W.E., Steam Turbine Over Speed Protection, *Chemical Processing*, Vol. 59, No. 7, 1996, pp. 48–54.

139. Nelson, W.R., Haney, L.N., Ostrom, L.T., Richards, R.E., Structured Methods for Identifying and Correcting Potential Human Errors in Space Operations, *Acta Astronautica*, Vol. 43, No. 3–6, 1998, pp. 211–222.

140. Nianfu Yang., B.S. Dhillon, Stochastic Analysis of a General Standby System with Constant Human Error and Arbitrary System Repair Rates, *Microelectronics and Reliability*, Vol. 35, No. 7, 1995, pp. 1037–1045.

141. Noda, H., A Soft-Error-Immune Maintenance-Free TCAM Architecture with Associated Embedded DRAM, *Proceedings of the IEEE Custom Integrated Circuits Conference*, 2005, pp. 451–454.

142. Norros, L., Human and Organisational Factors in the Reliability of Nondestructive Testing (NDT), *Proceedings of the Symposium on Finnish Research Programme on the Structural Integrity of Nuclear Power Plants*, 1998, pp. 271–280.

143. Nunn, R., Witt, S.A., Influence of Human Factors on the Safety of Aircraft Maintenance, *Proceedings of the International Air Safety Seminar*, 1997, pp. 211–221.

144. Nunn, R., Witts, S.A., The Influence of Human Factors on the Safety of Aircraft Maintenance, *Proceedings of the Flight Safety Foundation/International Federation of Airworthiness/Aviation Safety Conference*, 1997, pp. 212–221.

145. O'Connor, S.L., Bacchi, M., A Preliminary Taxonomy for Human Error Analysis in Civil Aircraft Maintenance Operations, *Proceedings of the Ninth International Symposium on Aviation Psychology*, 1997, pp. 1008–1013.

146. O'Leary, M., Chappell, S., Confidential Incidents Reporting Systems Create Vital Awareness of Safety Problems, *International Civil Aviation Organization (ICAO) Journal*, Vol. 51, 1996, pp. 11–13.

147. Oakhill, F., Human Factor in Maintenance, *Factory Management and Maintenance*, Vol. 94, No. 4, 1936, pp. 155–156.

148. Oldani, R.L., Maintenance and Inspection from the Manufacturer's Point of View, *Proceedings of the Meeting on Human Factors Issues in Aircraft Maintenance and Inspection*, 1989, pp. A17–A24.

149. Park, K.S., Jung, K.T., Estimating Human Error Probabilities from Paired Ratios, *Microelectronics and Reliability*, Vol. 36, No. 3, 1996, pp. 399–401.

150. Parker, J.F., A Human Factors Guide for Aviation Maintenance, *Proceedings of the Federal Aviation Administration Meeting on Human Factors Issues in Aircraft Maintenance and Inspection: Science, Technology, and Management: A Program Review*, 1992, pp. 207–220.

151. Pearl, A., Drury, C.G., Improving the Reliability of Maintenance Checklists, in Human Factors in Aviation Maintenance–Phase V: Progress Report, Office of Aviation Medicine, Federal Aviation Administration, Washington, D.C., 1995, pp. 127–165.

152. Pekka, P., Kari, L., Lasse, R., Study on Human Errors Related to NPP Maintenance Activities, *Proceedings of the IEEE Conference on Human Factors and Power Plants*, 1997, pp. 12/23–12/28.
153. Pekkarinen, A., Vayrynen, S., Tornberg, V., Maintenance Work during Shut-Downs in Process Industry: Ergonomic Aspects, *Proceedings of the International Ergonomics Association World Conference*, 1993, pp. 689–691.
154. Predmore, S.C., Werner, T., Maintenance Human Factors and Error Control, *Proceedings of the llth FAA/AAM Meeting on Human Factors in Aviation Maintenance and Inspection*, 1997, pp. 82–92.
155. Pyy, P., Laakso, K., Reiman, L., A Study on Human Errors Related to NPP Maintenance Activities, *Proceedings of the IEEE Conference on Human Factors and Power Plants*, 1997, pp. 12/23–28.
156. Ramalhoto, M.F., Research and Education in Reliability, Maintenance, Quality Control, Risk and Safety, *European Journal of Engineering Education*, Vol. 4, No. 3, 1999, pp. 233–237.
157. Raman, J.R., Gargett, A., Warner, D.C., Application of Hazop Techniques for Maintenance Safety on Offshore Installations, *Proceedings of the First International Conference on Health, Safety Environment in Oil and Gas Exploration and Production*, 1991, pp. 649–656.
158. Ramdass, R., Maintenance Error Management the Next Step at Continental Airlines, *Proceedings of the Flight Safety Foundation Annual International Air Safety Seminar*, 2005, pp. 115–124.
159. Ramdass, R., Maintenance Error Management, *Proceedings of the European Aviation Safety Seminar*, 2006, pp. 2–4.
160. Rankin, W., Hibit, R., Allen, J., Sargent, R., Development and Evaluation of the Maintenance Error Decision Aid (MEDA) Process, *International Journal of Industrial Ergonomics*, Vol. 26, 2000, pp. 261–276.
161. Rankin, W.L., The Maintenance Error Decision Aid (MEDA) Process, *Proceedings of the XIVth Triennial Congress of the International Ergonomics Association and 44th Annual Meeting of the Human Factors and Ergonomics Association*, 2000, pp. 795–798.
162. Rankin, W.L., User Feedback Regarding the Maintenance Error Decision Aid (MEDA) Process, *Proceedings of the International Air Safety Seminar*, 2001, pp. 117–124.
163. Rankin, W.L., Allen, J.P., Sargent, R.A., Maintenance Error Decision Aid: Progress Report, *Proceedings of the 11th FAA/AAM Meeting on Human Factors in Aviation Maintenance and Inspection*, 1997, pp. 19–23.
164. Rasmussen, J., Human Errors: A Taxonomy for Describing Human Malfunction in Industrial Installations, *Journal of Occupational Accidents*, Vol. 4, 1982, pp. 311–335.
165. Reason, J., *Human Error*, Cambridge University Press, Cambridge, UK, 1990.
166. Reason, J., Maddox, M.E., Human Error, in Human Factors Guide for Aviation Maintenance, Report, Office of Aviation Medicine, Federal Aviation Administration, Washington, D.C., 1996, pp. 14/1–14/45.
167. Reason, J., Approaches to Controlling Maintenance Error, *Proceedings of the 11th FAA/AAM Meeting on Human Factors in Aviation Maintenance and Inspection*, 1997, pp. 9–17.
168. Reason, J., Corporate Culture and Safety, *Proceedings of the Symposium on Corporate Culture and Transportation Safety*, April 1997, pp. 187–194.
169. Reason, J., Maintenance-Related Errors: The Biggest Threat to Aviation Safety after Gravity? *Proceedings of the International Aviation Safety Conference*, 1997, pp. 465–470.

170. Reason, J., *Cognitive Engineering in Aviation Domain*, Lawrence Erlbaum Associates, Mahwah, NJ, 2000.

171. Reason, J., Hobbs, A., *Managing Maintenance Error: A Practical Guide*, Ashgate Publishing Company, Aldershot, UK, 2003.

172. Reiman, L., Assessment of Dependence of Human Errors in Test and Maintenance Activities, *Proceedings of the International Conference Devoted to the Advancement of System-Based Methods for the Design and Operation of Technological Systems and Processes*, 1994, pp. 073/23–28.

173. Robertson, M.M., Using Participating Ergonomics to Design and Evaluate Human Factors Training Programs in Aviation Maintenance Operations Environments, *Proceedings of the XIXth Triennial Congress of the International Ergonomics Association*, 2000, pp. 692–695.

174. Rockoff, L.M., Anderson, D.E., Evelsizer, L.K., Human Factors in Aero Brake Design for EVA Assembly and Maintenance, *Society of Automotive Engineers (SAE) Transactions*, Vol. 100, No. 1–Part 2, 1991, pp. 1526–1536.

175. Rogan, E., Human Factors in Maintenance and Engineering, In Human Factors in Aviation Maintenance–Phase V: Progress Report, Office of Aviation Medicine, Federal Aviation Administration, Washington, D.C., 1995, pp. 255–259.

176. Rowekamp, M., Berg, H.-P., Reliability Data Collection for Fire Protection Features, *Kerntechnik*, Vol. 65, No. 2, 2000, pp. 102–107.

177. Russell, S.G., The Factors Influencing Human Errors in Military Aircraft Maintenance, *Proceedings of the International Conference on Human Interfaces in Control Room*, 1999, pp. 263–269.

178. Schmidt, J., Schmorrow, D., Figloc, R., Human Factors Analysis of Naval Aviation Maintenance Related Mishaps, *Proceedings of the XIVth Triennial Congress of the International Ergonomics Association*, 2000, pp. 775–778.

179. Schumacher, J.L., Maintenance Personnel Initiatives in Repair Stations, *Proceedings of the 4th Federal Aviation Administration Meeting on Human Factors Issues in Aircraft Maintenance and Inspection: Trends and Advances in Aviation Maintenance Operations*, 1994, pp. 63–74.

180. Segerman, A.M., Covariance as a Metric for Catalog Maintenance Error, *Proceedings of the AAS/AIAA Space Flight Meeting*, 2006, pp. 2109–2127.

181. Seminara, J.L., Human Factor Methods for Assessing and Enhancing Power Plant Maintainability, Report No. EPRI-NP-2360, Electric Power Research Institute, Palo Alto, California, 1982.

182. Seminara, J.L., Parsons, S.O., Human Factors Review of Power Plant Maintainability, Report No. EPRI-NP-1567, Electric Power Research Institute, Palo Alto, California, 1981.

183. Seminara, J.L., Parsons, S.O., Human Factors Engineering and Power Plant Maintenance, *Maintenance Management International*, Vol. 6, No. 1, 1985, pp. 33–71.

184. Shepherd, W.T., Human Factors in Aviation Maintenance—Eight Years of Evolving R&D, *Proceedings of the Ninth International Symposium on Aviation Psychology*, April 27–May 1, 1997, pp. 121–130.

185. Shepherd, W.T., A Program to Study Human Factors in Aircraft Maintenance and Inspection, *Proceedings of the Human Factors Society 34th Annual Meeting*, 1990, pp. 1168–1170.

186. Shepherd, W.T., Human Factors in Aircraft Maintenance and Inspection, *Proceedings of the International Conference on Aging Aircraft*, 1991, pp. 301–304.

187. Shepherd, W.T., Human Factors Challenges in Aircraft Maintenance, *Proceedings of the Human Factors Society 36th Annual Meeting*, 1992, pp. 82–86.

188. Shepherd, W.T., Human Factors in Aviation Maintenance: Program Overview, *Proceedings of the 7th Federal Aviation Administration Meeting on Human Factors Issues in Aircraft Maintenance and Inspection: Science, Technology, and Management: Program Review*, 1992, pp. 7–14.

189. Shepherd, W.T., Johnson, W.B., Human Factors in Aviation Maintenance and Inspection: Research Responding to Safety Demands of Industry, *Proceedings of the Human Factors and Ergonomics Society 39th Annual Meeting*, Vol. 1, 1995, pp. 61–65.

190. Shepherd, W.T., Kraus, D.C., Human Factors Training in the Aircraft Maintenance Environment, *Proceedings of the Human Factors and Ergonomics Society Meeting*, 1997, pp. 1152–1153.

191. Sola, R., Nunez, J., Torralba, B., An Overview of Human Factor Activities in CIEMAT, *Proceedings of the IEEE Sixth Annual Human Factors Meeting*, 1997, pp. 13/1–13/4.

192. Spray, W., Teplitz, C.J., Herner, A.E., Genet, R.M., A Model of Maintenance Decision Errors, *Proceedings of the Annual Reliability and Maintainability Symposium*, 1982, pp. 373–377.

193. Sridharan, V., Mohanavadivu, P., Reliability and Availability Analysis for Two Non-identical Unit Parallel Systems with Common Cause Failures and Human Errors, *Microelectronics and Reliability*, Vol. 37, No. 5, 1997, pp. 747–752.

194. Strauch, B., Sandler, C.E., Human Factors Considerations in Aviation Maintenance, *Proceedings of the Human Factors Society 28th Annual Meeting*, Vol. 2, 1984, pp. 913–915.

195. Su, K.W., Hwang, S.L., Liu, T.H., Knowledge Architecture and Framework Design for Preventing Human Error in Maintenance Tasks, *Expert Systems and Applications*, Vol. 19, No. 3, 2000, pp. 219–228.

196. Sung, C., Development of Optimal Production, Inspection, and Maintenance Strategies with Positive Inspection Time and Preventive Maintenance Error, *Journal of Statistics and Management Systems*, Vol. 8, No. 3, 2005, pp. 545–558.

197. Sur, B.N., Sarkar, T., Numerical Method of Reliability Evaluation of a Stand-by Redundant System, *Microelectronics and Reliability*, Vol. 36, No. 5, 1996, pp. 693–696.

198. Taylor, J.C., Organizational Context for Aircraft Maintenance and Inspection, *Proceedings of the Human Factors Society 34th Annual Meeting*, 1990, pp. 1176–1180.

199. Taylor, J.C., Reliability and Validity of the Maintenance Resources Management/Technical Operations Questionnaire, *International Journal of Industrial Ergonomics*, Vol. 26, 2000, pp. 217–230.

200. Taylor, J.C., The Evolution and Effectiveness of Maintenance Resource Management (MRM), *International Journal of Industrial Ergonomics*, Vol. 26, No. 2, 2000, pp. 201–215.

201. Toriizuka, T., Application of Performance Shaping Factor (PSF) for Work Improvement in Industrial Plant Maintenance Tasks, *International Journal of Industrial Ergonomics*, Vol. 28, No. 3–4, 2001, pp. 225–236.

202. Trotter, B., Maintenance and Inspection Issues in Aircraft Accidents/Incidents, *Proceedings of the Human Factors Issues in Aircraft Maintenance and Inspection*, 1989, pp. A6–A8.

203. Underwood, R.I., Occupational Health and Safety: Engineering the Work Environment—Safety Systems of Maintenance, *Proceedings of the International Mechanical Engineering Congress on Occupational Health and Safety*, 1991, pp. 5–9.

204. Varma, V., Maintenance Training Reduces Human Errors, *Power Engineering*, Vol. 100, No. 8, 1996, pp. 44, 46–47.
205. Vaurio, J.K., Optimization of Test and Maintenance Intervals Based on Risk and Cost, *Reliability Engineering and System Safety*, Vol. 49, No. 1, 1995, pp. 23–36.
206. Vaurio, J.K., Modelling and Quantification of Dependent Repeatable Human Errors in System Analysis and Risk Assessment, *Reliability Engineering and System Safety*, Vol. 71, No. 2, 2001, pp. 179–188.
207. Veioth, E.S., Kanki, B.G., Identifying Human Factors Issues in Aircraft Maintenance Operations, *Proceedings of the Human Factors and Ergonomics Society Meeting*, 1995, pp. 950–951.
208. Vlenner, C.A., Drury, C.G., Analyzing Human Error in Aircraft Ground Damage Incidents, *International Journal of Industrial Ergonomics*, Vol. 26, No. 2, 2000, pp. 177–199.
209. Walter, D., Competency-Based On-the-Job Training for Aviation Maintenance and Inspection: A Human Factors Approach, *International Journal of Industrial Ergonomics*, Vol. 26, No. 2, 2000, pp. 249–259.
210. Wanders, H.J.D., Improving Production Control Through Action Research: A Case Study, *Maintenance Management International*, Vol. 6, No. 1, 1985, pp. 23–31.
211. Wang, C.H., Sheu, S.H., Determining the Optimal Production-Maintenance Policy with Inspection Errors: Using a Markov Chain, *Computers and Operations Research*, Vol. 30, No. 1, 2003, pp. 1–17.
212. Ward, M., McDonald, N., An European Approach to the Integrated Management of Human Factors in Aircraft Maintenance Introducing the IMMS, *Proceedings of the Conference on Engineering Psychology and Cognitive Ergonomics*, 2007, pp. 852–859.
213. Wenner, C.L., Drury, C.G., A Unified Incident Reporting System for Maintenance Facilities, In Human Factors in Aviation Maintenance–Phase VI: Progress Report, Vol. II, Office of Aviation Medicine, Federal Aviation Administration, Washington, D.C., 1996, pp. 191–242.
214. Wenner, C.L., Wenner, F., Drury, C.G., Spencer, F., Beyond "Hits" and "Misses": Evaluating Inspection Performance of Regional Airline Inspectors, *Proceedings of the 41st Annual Human Factors and Ergonomics Society Meeting*, 1997, pp. 579–583.
215. Wen-Ying, W., Der-Juinn, H., Yan-Chun, C., Optimal Production and Inspection Strategy While Considering Preventive Maintenance Errors and Minimal Repair, *Journal of Information and Optimization Sciences*, Vol. 27, No. 3, 2006, pp. 679–694.
216. Williams, J.C., Willey, J., Quantification of Human Error in Maintenance for Process Plant Probabilistic Risk Assessment, *Proceedings of the Institution of Chemical Engineers Symposium*, 1985, pp. 353–365.
217. Winterton, J., Human Factors in Maintenance Work in the British Coal Industry, *Proceedings of the 11th Advances in Reliability Technology Symposium*, 1990, pp. 312–322.

Index

A

Absorption Law, 14
Acceleration, vibration, 32
Accident, 4
Adjust (task), 64
Aerojet General Method, 41
Aircraft
  accidents, maintenance errors, 3, 99–100
  classification, maintenance challenges, 80
  maintenance guidelines, 107–109
  maneuvering, time-continuous tasks, 37
Air Force Inspection and Safety Center Life
    Sciences Accident and Incident Reporting
    System, 41
Air Midwest Raytheon (Beechcraft) 1900D
    accident, 109
Align (task), 64
Amplitude, vibration, 32
AND logic gate, 53, 55–56
Angle viewing, *see* Sight
Anthropometrics, 92
Asoka, of India, 13
Assembly
  human errors, 40
  sequence, aviation maintenance, 100
Associative Law, 13–14
Auditory systems, *see* Noise
Automated systems, categories, 30
"Automation" (FAA human factors guidebook
    chapter), 82
Availability, power generation plant goal, 91
Average human reliability, 3
Aviation maintenance
  Air Midwest Raytheon (Beechcraft) 1900D
    accident, 109
  British Airways BAC1-11 accident, 110
  case studies, 109–110
  categories of, 100
  cause-and-effect diagram, 101–102
  causes of, 100
  challenges, 80
  common errors, 101
  common problems, 86
  Continental Express Embraer 120 accident,
    109
  error analysis methods, 101–106
  error-cause removal program, 102–104
  examples, 99–100
  facts and figures, 99–100
  fault tree analysis, 104–106

  fundamentals, 79, 99
  guidelines, 80–83, 107–109
  impact of human factors, 79–80
  integrated maintenance management system,
    83–84
  maintenance error decision aid, 106–107
  problems, common, 86
  training, 84–86
  types and frequency, 100–101
Aviation Safety Reporting System, 41

B

Behavior factors, 129
Bell Telephone Laboratories, 45, 53
Benefits, human factors engineering applications,
    96
Binomial distribution, 21
Boeing aircraft
  case study, 100
  737-200 structural failure, 4, 100
  747 structural failure, 3
Boiling water reactor nuclear power plant, 113
Books, informational, 5–6
Boolean algebra laws, 13–16
  Absorption Law, 14
  Associative Law, 13–14
  Commutative Law, 14
  Distributive Law, 14
  Idempotent Law, 14
Bricklaying, study of, 1
British Airways BAC1-11 accident, 110
Brooks Air Force Base, 1
Bunker-Ramo Tables, 41

C

Case studies
  aviation maintenance, 109–110
  power plant maintenance, 92–93
Categories, *see* Classifications and categories
Cause-and-effect diagram (CAED), 49–50,
    101–102
Causes, aviation maintenance errors, 100
Central Research Institute of Electric Power
    Industry (CRIEPI), 116
Challenges, aviation maintenance, 80
Character height estimation formula, 33
Clapham Junction railways accident, 3, 63